BEDROCK GEOLOGY

Patrick Nurre

The Northwest Treasures Curriculum Project
Building Faith for a Lifetime of Faith

BEDROCK GEOLOGY

By Patrick Nurre

Bedrock Geology
Published by Northwest Treasures
Bothell, Washington
425-488-6848
NorthwestRockAndFossil.com
northwestexpedition@msn.com
Copyright 2013 by Patrick Nurre.
All rights reserved.
ISBN-13: 978-1495206627
ISBN-10:1495206629

BEDROCK GEOLOGY

Introduction

When I have been teaching a straight forward interpretation of Genesis, there are certain questions that normally arise, usually from people who have trusted what secular science says about the age of the Earth. These questions include, "Why is it necessary to get the Book of Genesis right?" "Why are the things in Genesis an issue?" "Why make a big deal out of seemingly side issues like young Earth or old Earth or a Global Flood? After all, they are not critical to salvation."

These are legitimate and very important questions. Here are some things to consider:

1. If the Bible is the word of God, not just the reasoning of man or the philosophy of man, then there is an issue of **authority**. Faith and obedience are important matters for the believer in Christ. At issue is whether God has spoken to man and if so, will he respond accordingly?

2. If the Bible is truth, not just good moral ideas, then there is an issue of **integrity**. Can God's words be understood and do they mean what they say? Can we trust it if it is wrong when it comes to understanding Earth history?

3. If the Bible is recorded history, then there is an issue of **accuracy**. Do God's words communicate what actually took place in history, or are the words just great moral stories? If they are just great moral stories, then there is no reality in history. If there is no reality in history, how do we know these stories are even true and trustworthy?

4. If the Bible is a record of God's will and plan for man, then there is an issue of **authenticity**. Is Jesus the Messiah, the Promised One, the One who will fix the issue of sin and salvation, or is He just another great moral teacher?

Consider the following:

- **Jesus is called the Son of David.** God had promised David that one of his descendants would sit on his throne forever. In order to demonstrate this, the genealogy of Jesus must be true and accurate. Did He really descend from David? If His genealogy back to David (Old Testament) is inaccurate, then Jesus cannot be the fulfillment of the promise made to David. The Old Testament must be accurate at this point. It cannot be wrong about one thing about the genealogy between David and Jesus, or there will be a question as to the fulfillment of the promise through Jesus.

- **Jesus is called the Lion of Judah.** Jesus must be able to show that He came from the tribe of Judah. Judah was the son of Jacob. Can this be shown to be true or not? Is the genealogy of Jesus accurate all the way back to Judah? The Old Testament must be accurate at this point. If even one thing in Genesis, initially where Jacob appears, is wrong, then there is a question if even Judah was a real person and the legitimacy of Jesus is at stake.

- **Jesus is claimed to be a descendant of Abraham in the Gospels.** Why is this important? Abraham was given a promise by God that one of his descendants would be the blessing for the entire Earth. Abraham begins his life in the Book of Genesis. Is Genesis completely accurate in all it states or not? Is the genealogy of Jesus all the way back to Abraham accurate? If Genesis is not trustworthy in all it says, then there is even a question as to whether God ever spoke to Abraham and gave him a promise. The Old Testament must be accurate in all it says or the rest of the Bible is suspect.

- **In Romans 5:12-15** Paul describes three historical events from the Book of Genesis that if they could be shown to be false, the entire Bible would fall to pieces. One could not even claim that the Bible was filled with good moral principles. If the Scriptures are inaccurate in one place, then it will be suspect in all places. How would we ever know the real truth? Therefore we would not be able to trust it in any place. Here are those historical events.
  (1) Sin entered the world through one man, Adam.
  (2) Death came as a result of Adam's one sin.
  (3) Death and sin spread to all men (descendants of Adam), and this was shown to be the case because all sinned from Adam onward.
  So, Genesis must be accurate in all it states about Adam and his origin or we must discount the entire New Testament when it talks about sin, consequences for our sin and our need for a savior. We must also discount what the Bible states about Jesus being that savior because after all, how do we know anything that the Bible states, is true? If we cannot trust the Scriptures about creation and the flood, how can anything the New Testament teaches be trusted?

- **Jesus and Peter claim that the global flood of Genesis was a warning that men are accountable to God for all they do and say. The New Testament also teaches that the flood is a foreshadow of the last judgment.** If the flood did not happen the way Genesis says it did, then what Jesus said is simply not true or at best is a fictitious story. Further, Peter would also be false. The New Testament would be whittled down to almost nothing if the writings of Peter, Paul and the Gospels cannot be trusted. If the Flood did not historically happen, then all references to judgment, sin, condemnation, penalties, guilt, and rebellion are cruel and should be removed.

The entire foundation of all that Christians believe rests on what happened in Genesis. At issue is whether it is 100% true and accurate or not? If not, then Christians are without hope and the world is without hope because nothing else has been revealed about God and what He requires. The whole idea of a God who created us is questionable.

So, yes, studying the Book of Genesis as history and truth is critical when approaching the record of creation and a global flood. This is why building a proper framework and worldview for interpreting the Earth around us is critical. And yes, it does matter whether the Earth is young or old because the Bible categorically teaches that it is young. If the Bible is wrong on this point, then it is wrong, or at best is suspect, on every other point it makes.

One final point to consider – according to the genealogies in Genesis, each father and son relationship that Genesis records is listed in order and the son's birth at a certain age of the father. Only a straight forward interpretation of this is possible. If there are names missing from the list that establish the connections from Adam to Abraham, then the genealogies are worthless. They don't establish anything including a trustworthy connection from Adam to Abraham.

I am related to William the Conqueror. To say I am related to William the Conqueror may sound impressive, but there are hundreds of names that must be furnished in order to verify my claim. It takes a generation by generation connection to establish my claim, or I have no claim. It becomes more important when truth and eternity are at stake.

If the Creation/Fall and Flood stories are not historically true, then we have a problem in establishing the connection from Adam to Abraham. Both of these historical records are absolutely necessary to get right in order to assure their veracity. The genealogies make it apparent that among other things, God wanted crucial stories to be remembered. There is an unbroken link between Adam and Abraham, but only if the genealogies and the Creation/Fall/Flood stories are true and correct. Adam could have known Lamech, the father of Noah. Lamech knew Noah's sons, Shem, Ham and Japheth. Shem could have known Abraham. Abraham knew Isaac and could have known Jacob. These 7 connecting points to Adam would have assured an unbroken story of man's Creation, the Fall and the Flood – three crucial historical events in the life of man! Are they real historical events or just mythical stories?

Patrick Nurre
May 31, 2013
Northwestexpedition@msn.com

BEDROCK GEOLOGY
The Northwest Treasures Curriculum Project

Description of the Curriculum Project
To the Parent or Guardian

The passion and goal of The Northwest Treasures Curriculum Project is to inform, educate, equip and inspire confidence in the historical record we call the Scriptures and to provide affordable resources that will help achieve this. The uniqueness of this Project lies in its presentation. Through my experience with children and with geology textbooks, I have come to the conclusion that we as parents tend to major on the minors and minor on the majors when it comes to geology. Many of us consider geology to be an intimidating subject that we know so little about. Some consider it to be a "necessary evil"; something to check off our list of subjects to be taught before our child graduates. For me Earth Science class was an absolute bore as a kid. The teacher did not want to teach it. I did not want to learn it and consequently I got into more trouble in my 7th grade Earth Science class than in any other class in school. So little time is spent connecting the study of Earth History and all its various parts into a complete worldview. It is generally taught once in the child's schooling experience because he/she has to have so many "credits" of science. Very little time is spent actually observing the Earth and handling the rocks and fossils, and yet most children love rocks, minerals, volcanoes and dinosaurs. The subject is a natural for most kids and it is here that we have a wonderful opportunity to prepare our children both to believe in what God has done in the past and to equip them for their futures.

The Curriculum Project consists of a foundational geology course (*Bedrock Geology*) for grades 5-12 (and includes a special section for PreK-4th) where the basics of geology will be presented from a Biblical (Scriptural) perspective. *Bedrock Geology* is like the hub of a wheel. From this hub will emanate spokes that are the various Northwest Treasures unit studies in geology that you can do after completing *Bedrock Geology*. They will help flesh out the various components of Earth history. The idea behind this is to provide a concise and affordable Earth science curriculum designed to promote a Biblical worldview all the way through the child's schooling experience so that he/she will have a sound view of Earth history from God's perspective and will be able to defend that position. *Bedrock Geology* is not meant to be an exhaustive geology course in itself, but to provide a framework, built on the Scriptures, through which to interpret the various rock layers and landforms of the Earth.

Bedrock Geology will include these four basic subjects. Completion of the text/activities with quizzes and exam would constitute one semester of Earth Science credit.

I. **The History of Modern Geology** – When, where and how our culture has developed its modern understanding of Earth history and where it has conflicted with the Scriptures

II. **The Origin and Nature of the Earth** – The order and cause behind the Earth on which we live; the atmosphere, water supply, igneous rocks, and minerals all around us taking a close look at the Book of Genesis as the interpretive framework

III. **The Genesis Flood** – The historical turning point in Earth history explored as an actual historical event with geological consequences we still observe and experience today; also a discussion of metamorphic and sedimentary rocks and fossils examined through the interpretive framework of the Book of Genesis

IV. **The Ice Age** – The aftermath of the Genesis Flood; its causes and effects on our Earth immediately stemming from the Genesis Flood. This section is one of the least understood and least studied for the typical child. And most Christians view any study of an ice age with suspicion. I will help unravel the mystery and show how it is possible to explain an Ice Age within the framework of Genesis

Each one of these themes will be explored both within the Bedrock Geology and through the various unit studies available from Northwest Treasures. It is my belief that geology is the most important subject of any child's schooling experience. This is because it is a comprehensive worldview that affects so many outcomes in his/her lifetime. The greatest wedge in the church today when it comes to studying Genesis (the Book of beginnings), is the issue of the age of the Earth. This is a geological issue as well as a spiritual issue. What a person believes about Genesis, origins, young earth/old earth and a global flood will affect the rest of his/her spiritual health and life. I believe some aspect of Earth history must be taught each semester to ensure a complete worldview. It is important and geology should not be taught as an isolated subject to check it off our list of school requirements.

Basic Materials Needed:

(Some of these are provided in the *Bedrock Geology* curriculum kit – **bold lettered**):

- *Bedrock Geology* **text**
- <u>Genesis – Rock Solid</u> by Patrick Nurre (Supplemental reading)
- Pencils
- Colored Pencils
- Ruler
- A Bible
- Lined notebook (spiral or 3 ring binder is acceptable); the student will be recording his/her answers to activities, vocabulary words, and notes from discussions
- Magnifying Glass
- A basic dictionary, like Merriam-Webster's
- Access to Encyclopedias or the Internet
- **A Basic Rocks, Minerals and Fossils Kit**
- Access to basic kitchen supplies generally common to all kitchens

Secondary Materials (Optional, but Encouraged):

A variety of unit studies are available from Northwest Treasures that will help cement the difference between the modern view of geology and the Biblical geology view. It is hoped that you will promote further study by incorporating these kits at least once per semester. The Northwest Treasures Curriculum Project is set up this way to facilitate your budget. The study of geology is a critical component of your child's education. Do not underestimate this. There are a variety of suggested kits and books in Appendix F.

Suggestions for Using this Curriculum:

1. Suggestion for teaching the material – *Bedrock Geology* is broken up into 4 parts. Tackle each part as a unit. Have your child *memorize* the passages of Scripture. One of the biggest problems for students learning Biblical Geology, is that they just do not know their Scriptures. They have been exposed to the Uniformitarian view with such regularity, that this becomes a sort of fall-back position. To learn the Biblical view, we must learn the Scriptures. In addition to knowing the Scriptures, your child must also learn the "language" of geology. At the start of each Section is a list of geological terms. I would suggest devoting a section of the student notebook to these words, and learning them before doing the study. They will help him or her find their way around the subject of geology.

A very basic suggestion for each week:

Day 1: Vocabulary study (create a glossary in your notebook); read text.
Day 2: Read text again; Hi-light important ideas; discuss the section; do activities.
Day 3: If necessary, finish activities. Take Quiz if indicated. There is a Final Comprehensive Exam.

2. Remember that your goal is to build a complete worldview into your child. So the time spent using and discussing this material should facilitate that goal. This curricula can be covered in one semester, with about one section per week. I would suggest formal class time of no less than 2 times per week. Any less and the student will not remember what you discussed from the previous lessons. Likewise, to quickly cover the material in order to get done will not ensure that the student is effectively learning. You should take as long as is necessary to truly learn the concepts.

3. Make sure you are reinforcing what the child is learning by asking questions as you are out and around or while the child is watching TV or playing video games. You can ask questions like, "How does that program or game reinforce what we talked about?" "How does that advertisement tear down what you learned about a Biblical worldview?" "Why do Mom and Dad obey the speed limit?" etc.

4. Pay attention to the books your child is reading or checking out of the public library. Again, you can reinforce what you are talking about by asking questions.

5. Always be mindful of behavior – yours and your child's. Behavior that is contradictory to the Scriptures will tear down a Biblical worldview as easily as will paying attention to the wrong information.

6. If you desire your PreK-4th grader to follow along with this study, they should begin with <u>An Introduction to Biblical Geology</u>, located in Appendix B, and then move on to the unit studies such as:

Noah's Flood for Grades PreK-2nd   Rocks and Minerals for Grades PreK-2nd
Dinosaurs for Grades 3-6   Volcanoes for Grades PreK-2nd
Fossils for Kids   <u>God's Design for Heaven and Earth</u>
Creation Science (Grades K-3)

These are just ideas of possible unit studies you could do. There are many other choices available. See *NorthwestRockAndFossil.com* or Appendix F for other suggestions. These kits are made by us and available exclusively from *Northwest Treasures*.

Niagara Falls – During his visit to Niagara Falls in 1841 Charles Lyell formulated his age for the Falls. His explanation, published in England after he returned, has served as a major reason for rejecting the Bible's chronology and ruining faith in the Scriptures. What Lyell failed to report were the conflicting reports of the rates of erosion of the Falls, some of which lined up with Biblical chronology. Had he been honest and mentioned these, the history of modern geology might have been a different one from what has taken place.

I The History of Modern Geology
Section 1: The Worldview

Vocabulary: *Write out a brief description for each word and name in your notebook: (Not all of these words are defined in your text. You will need to look up many of them in a dictionary.)*

Geology   Science   Uniformitarianism   Worldview   Erosion
Naturalism (or naturalistic)   Stratigraphy   Nicolas Steno
Antiquated   Geologic Time Table (The Geologic Column)   Myth   Bias

Teacher Note: For grades PreK-4th the ideas found in *An Introduction to Biblical Geology* in Appendix B should help you distill the points in the first section.

Until the 1700's, the study of geology included the Bible as a necessary textbook, and was viewed as the correct historical explanation of how the Earth was created as well as the rock layers being formed by the Genesis Flood. Today, the predominant view, and the only one that is accepted by scientists, of Earth science (*geology* or *Earth history*) does not include any reference to the Bible. The Bible is viewed as *antiquated* and *mythological*. How did we arrive at our modern view of Earth history? Was the world blinded for thousands of years until modern geology shed light on the subject and delivered us from this ancient blindness?

Actually the story is much more complicated than this and it involved a change in the way man thinks about himself, God and the world around him. Earth science involves so much more than the study of rocks and minerals. It is also a worldview; a framework; a human interpretation of the Earth, and how it came to be the way it is.

For over 200 years the history of the Earth has been being rewritten. Until the late 1700s the study of the Earth was done in light of the Scriptural account of Creation and the Genesis Flood contained in Genesis chapters 1-9. For example *Nicolas Steno* (1638-1686) was a geologist who believed in the Biblical revelation of a 6 day creation and that the rock layers and fossils were the result of the global flood of Genesis chapters 7-9. He is considered to be one of the founders of the science of *stratigraphy*. His principles are still in use today. What happened to change this view?

Nicolas Steno (Danish: Niels Stensen;
Latinized to Nicolaus Stenonis or Nicolaus Stenonius)

The modern understanding of Earth history has been created by speculation and belief fueled by blindness or indifference to the spiritual realm and to Biblical history. Many geologists have ignored these, thinking that they have nothing to do with the way they view the past history of our Earth. Because of this, most geologists have developed a lopsided view of Earth history. This view is actually a philosophy called **uniformitarianism** or *"The present is the key to the past"*. In other words geologists think they can gain an understanding about Earth history by only studying present geological forces and processes. We will discuss this more in a later section.

What makes the modern geological view confusing for many is that while it incorporates true science that involves studying the nature of rocks, minerals, elements and mountains, modern geology also involves ***human interpretation*** about the *origin* of those things. This is too often not made clear, and the student learns geology as a unified scientific package. For example, we often cannot see the difference between the ***fact*** of erosion and the ***application*** of erosion to the past history of our Earth. One involves studying the present process of erosion and its present effect. For example, *"A heavy rain that dropped 12 inches in a four hour period produced this washout of a local neighborhood"*. The other involves an application of erosion to the past based on the uniformitarian belief system. *"A river at the bottom of canyon wears away at the sides of the canyon. Therefore the river has been cutting the canyon for millions of years."*

Some geologists will state that they are like detectives, that their job is to piece together the history of the Earth much like a detective solves a crime. They gather evidence and then formulate conclusions. But the conclusions are often the result of personal interpretations formed by their personal beliefs, which means excluding any explanation that includes the Bible. For instance, the age of the Earth according to uniformitarian principles, has been constantly expanding from a few thousand years old in the 1700s to over four billion years old in the current century. Geologists will insist this is because their understanding has grown through scientific research. But this picture is not an accurate one. To ignore vital information is to skew the results and so present a historical picture that is untrue or is at best incomplete. Geologists should be like detectives. The good detective should draw from as many sources of information as possible.

So, is modern geology really the result of unbiased scientific research and good detective work? Quite often, the geologist of today is totally unaware of two vitally important parts in his/her understanding and formulating Earth history: **(1)** Many geologists do not recognize the influence of their own *worldview* or biases in formulating their descriptions about the past history of the Earth. **(2)** Virtually all geologists ignore the Scriptural historical account of the creation and global flood.

Why incorporate the Bible? Isn't the Bible just a collection of myths and therefore not suitable in the study of geology? Actually the Bible has been repeatedly recognized as a historically accurate account of our past. For example David Rohl, Egyptologist and Archaeologist in his book, <u>Pharaohs and Kings, A Biblical Quest</u>, lists 40 archaeological discoveries concerning Abraham, Joseph, Moses, David and Solomon that verify the ancient history presented in the bible. David Rohl claims to be an agnostic but interestingly also believes that the Biblical history is accurate. The Bible's history has been routinely rejected when researching man's history. Quite the contrary, the Bible's historical account should be a vital part of interpreting Earth history!

Geologists fail to realize that uniformitarianism, the foundation of modern geology, is also an attempt to tell a historical story of Earth and man. It is presented as pure science, but it is not pure science, but an alternative view of Earth history. The Biblical view of geology as history records how God has worked in the world He created. The uniformitarian view leaves this valuable information out.

In addition to being an alternative historical view of Earth history, uniformitarianism is also a spiritual view of Earth history. It is atheistic and naturalistic. Naturalism is a worldview. It is a way of looking at the world. It is a belief that excludes any possibility of a God who was and is involved in the creation. In the modern age science has become synonymous with naturalism. However, these two disciplines are mutually exclusive. Pure science is drawing conclusions about what is observed and measured. It is not naturalism, atheism or theism. Atheism, theism and naturalism are beliefs, not science. But modern scientists insist on the absence of God.

The Bible in contrast is theistic and supernaturalistic. Both naturalism and supernaturalism are views that cannot be demonstrated by scientific research to be true. But modern geology fails to identify this when telling its view of Earth history to the public. Modern geology and Biblical geology are diametrically opposite *beliefs* and equally as *spiritual*. And both are attempts to tell a story of Earth history.

Because of its view that the Bible as mythical history and not scientific, modern geology leaves out a vital part of studying Earth history – the Scriptures, which have been shown to be a reliable record of the history of the past. The Scriptures should therefore be considered among the various sources of knowledge when studying Earth history. A detective that leaves out vital information in solving a crime will undoubtedly render a biased conclusion. If he is convinced from the very beginning about the nature of the crime and who committed it, all his research will be slanted toward his bias. This will produce a great deal of misunderstanding.

Such has been the history and development of modern geology. It is a combination of science, religion or belief about the past, and history. That's right. Earth history is after all, about HISTORY! It is about what took place in the past. Despite the confidence that modern geologists have about their interpretation of Earth history, their ability to provide suitable details has fallen short. This is precisely why Earth history as presented by modern geology is not purely science.

Activity A: Record your answers to the following questions in your notebook. Answers can be found in the Activity Answer Key.
   1. *Who was Nicolas Steno?*
   2. *What is Naturalism?*
   3. *What is Uniformitarianism?*
   4. *What is science? What in modern geology is science, and what is not?*
   5. *How is modern geology like the Biblical view of Creation and the Flood?*

Cross-sectional view of lava flows of the Columbia River flood basalts, part of the lava plains of the western United States. Were these laid down rapidly in quick succession or over hundreds of thousands or even millions of years?

Friedrich Wilhelm Nietzsche (1844-1900) – a child of The Enlightenment who did much damage to the acceptance of Christian faith with his pronouncement that, "God is dead," and by advancing the ideas of Evolution and the glory of man.

I The History of Modern Geology
Section 2: The Movers and Shakers

Vocabulary: *Write out a brief description for each word and name in your notebook:*

The Enlightenment	James Hutton	Charles Lyell	Deism
Catastrophism	Atheism	Theism	
Framework			

The distinction of the title, Father of Modern Geology, has been given to **James Hutton,** a Scottish physician (1726-1797) who lived during the period of history known as **The Enlightenment.** It was James Hutton who radically broke from the tradition of viewing the Earth's rock layers in light of the Genesis Flood. In 1797 Hutton wrote, *"The past history of our globe must be explained by what can be seen to be happening now. No powers are to be employed that are not natural to the globe, no action to be admitted except those of which we know the principle."*

Activity B: *In your own words, describe the problem with the previous statement. Is this a scientific statement? If not, why not? Is it good detective work? If not, why not? Record your answers in your notebook. Answers in Activity Answer Key.*

Hutton as painted by Sir Henry Raeburn

Was there something in the rock layers that led Hutton to formulate this radical departure from the Scriptural view? Rock layers are simply that – rock layers. They do not in themselves tell the whole story of how they got that way. A worldview, personal perspective or belief must serve as the framework in which to formulate ideas about the *"how"* or *"why"* of rock layers. It is here that modern geology fails to realize that this is exactly what their understanding of Earth history has been built on – a worldview; an interpretation! Hutton thought he saw evidence to support his view that the rock layers took a long time to become that way, but his conclusion was born simply from choosing to exclude that which he did not view as reliable, despite evidence to the contrary. His worldview was uniformitarian and naturalistic, not Biblical.

What was the predominate worldview in the days in which James Hutton lived and from where did it come from? The time in which James Hutton lived was a period of history in Western Culture called *The Enlightenment.* This period was itself a radical shift in the way men and women would look at God, the Bible, the church, and the Earth. No longer would the primary thinkers of the day look to the Scriptures for answers to life's perplexing questions about origins, the nature of the Earth and Mankind. Man would from this period on strive to be independent from any influence of the Scriptures and its record of God's involvement with His creation. It was not quite *atheism (there is no God),* but a radical departure nevertheless from viewing the Scriptures as inspired revelation from God.

In brief **the Enlightenment produced the following consequences:**

1. **A belief that man, apart from any biblical revelation, could discover his past**
2. **A belief that the Bible is at worst a myth and at best a corrupted record of the past**
3. **Gave birth to the religion of Deism; God created but is not now involved - God is therefore irrelevant in the scientific discussion of the history and nature of our world**
4. **The idea that science and "scientific" discovery alone could unravel our past**

Thus a new religion was born, the religion of **Deism** - God probably had created everything in the universe, but He was not now involved in it. The Earth and its history was a product of natural processes that had been going on for a long, long time. James Hutton was a Deist.

This brings me to the heart of the issue for the Christian. Although the subject of science (knowledge) in and of itself should only incorporate what can be observed and measured, it is clear from Scripture (Romans chapter 1) that knowledge about the created world around us should lead to the conclusion that there is a personal, absolute Creator to Whom we are accountable. But the scientific establishment refuses to incorporate anything to do with God because He is not science. This is a dilemma for the Christian who is subject to the Scriptures as his/her absolute rule of authority. The average Christian thinks he/she must compromise the Scriptures to be in harmony with modern science. The real solution to this is to demonstrate how modern science (geology) is another religion in disguise. Deism became the cloak of this disguise during the 1800s. To say that God was not involved in His creation subsequent to creating it, should have been unthinkable to the Christian during this time. All throughout the Scriptures we learn that God has been involved, continues to be involved and will be involved in His creation until the world comes to an end. This teaching is a central part of Biblical theology. Part of studying Biblical Geology is to become equipped to demonstrate this truth by showing how especially the Book of Genesis explains the various landforms and rocks all around us.

Out of Deism came the idea of uniformitarianism. This idea was to be popularized by another man, **Charles Lyell** (1797-1875). Although Lyell did not know Hutton personally, he was nevertheless influenced by him and by the current Enlightenment thinking of his day. Lyell also was a Deist. Lyell's personal goal in geology was to remove any influence of the Scriptural idea of a *catastrophic* global flood from the study of geology. He, more than any other, brought the idea of uniformitarianism into his culture of the day with the expression, "The present is the key to the past". Very simple and concise, wouldn't you agree? And the rest, as they say, is history! Today's geology is rooted in uniformitarianism. This idea continues to be the overriding principle that governs modern geologic thinking.

The elements of uniformitarian geology are:
1. The history of the Earth has been a totally naturalistic one
2. The history of the Earth has been a long, and at times a slow plodding of naturalistic processes
3. The history of the Earth has been punctuated by local catastrophes not to be confused with the global catastrophe of Noah's Flood
4. The Scriptures are to be excluded from any consideration of Earth history because they involve the idea of God. Since the idea of God is not testable by laboratory experiments, the idea is not science. Science alone holds the key to understanding Earth history
5. Only statements formulated by and in the context of science are to be permitted in the discussion of Earth history
6. The Geologic Timetable or Geologic Column, itself a result of uniformitarian thinking, is a man-made construct that interprets all geological data

Sir Charles Lyell

Activity C: Be sure to write your answers in your notebook. *Answers in the Activity Answers Key.*

1. In your own words describe how uniformitarianism differs from Genesis chapters 1 & 7
2. Describe how uniformitarianism is scientific; how it is religious
3. What are the challenges a believer in the Scriptures, as history, faces as he/she encounters modern uniformitarian geology?
4. In what ways does modern geology confuse the issue of a young earth view versus an old earth view of Earth history?

Read 2 Peter 3:3-8.

This passage of Scripture seems almost prophetic as it describes a philosophy about the Earth that will be prevalent among mankind. That philosophy is naturalistism. It is the belief that the history of the Earth has been going on for a long, long time with nothing different than has always been the case. Does this sound familiar? It should because it is describing uniformitarianism. That is the logical conclusion when God is rejected. All of Earth' processes have been going on as usual and naturalistically without the interference of a god.

Activity D: *The following is a chart comparing the two approaches to Earth history. Take some time and discuss it with another person.*

Genesis – Biblical Geology	Uniformitarianism – Secular Geology
Theism (God, but a specific God, the God of the Scriptures) – God has been actively involved in the origin and history of the Earth from the beginning; we call God's involvement in His creation, "miracles"	Deism (the deity; not identified, but sometimes referred to as "the Supreme Architect") – the Deity may have created the earth and universe but if He did, He is no longer involved in it; He does not intervene. Only natural laws govern the history of the Earth; there are no such things as miracles
Gen. 1:1 – in the beginning, **God**; "beginning" would mean of time and of matter; God is eternal; He did not have a beginning. Another way to say this is "in the beginning of our cosmos, God was already present. And it is He who has given order and meaning to creation.	**Big Bang** – in the beginning **matter** and space were already there; the Big Bang does not attempt to answer the question, "Where did matter and space come from?" It is ignored.
Gen. 1:1 – God created the **space** and the **earth** on Day 1 of the creation (created means, "out of nothing"); Earth was the only "heavenly body" in the beginning.	Matter exploded, **forming** galaxies, stars, and planets 15 billion years ago; Earth came along about 4.6 billion years ago or 11 billion years after the universe came into existence.
Gen. 1:2 – the earth was originally a surging mass of **water** (the "deep").	The earth was originally a cloud of **gas** or a **molten** blob of magma/lava; water came last
Gen. 1:2 – the **Spirit of God** was an active part of creation from the beginning.	God, spirit and religion are not matter and consequently are not relevant and are excluded from the study of Earth history
Gen. 1:3 – **light** (energy; the physical laws that govern the universe)	Physical laws and energy **evolved** as the universe evolved
Gen. 1:5 – "good" = perfect, done, complete in God's eyes; evening and morning define a complete revolution of the earth = day; **"one day"** means there was one complete 24 hour period of Earth history with evening and morning (time divisions) and it was the first day of more to come	"Day" in the Scriptures is simply **symbolical** for a period of time; because of the undeveloped minds of these early people, cosmology had to be communicated simply; they could not comprehend "deep time"; according to evolution, nothing is ever completed, but always changing into the next thing over millions of years ("deep time")

Take Quiz #1, p. 101.

This celebrated picture of Earth from space was taken in 1972 by the Apollo 17 crew while traveling to the Moon

II The Origin and Nature of the Earth
Section 1: In the Beginning

Vocabulary: *Write out a brief description for each word and name:*

Origins	Genesis	Evolution (Evolutionism)	Creation(Creationism)
Water	Atmosphere	Element	Hydrogen
Mineral	Atom	Big Bang	Rock
Cosmology	Molecule	Oxygen	Transcendence

In this section you will begin to develop a different worldview of the origin of the Earth. Very few Christians understand or even know their own history of the Earth and the life on it as presented in the Scriptures. They usually have some vague idea, but that idea has generally been developed within the worldview of uniformitarianism so commonly taught in our schools and in the library books we enjoy. The worldview present in the following Scriptures will begin to give you a new framework in which to interpret the landforms and the rocks on our planet.

Activity E: Read the following passages of Scripture: Genesis chapter 1

- *List the various things of creation that were accomplished on each day of creation week.*

- *Answer these questions: (1) On what day were the sun, moon and stars created? (2) On what day were the birds created? (3) On what day were all sea creatures created? (4) On what day were all land and air breathing animals created? (5) On what day was man created? (6) How does this order differ from the modern evolutionary view of life?*
    Answers in the Answers key.

Ever since astronomers have been able to gaze at the stars and other planets through their telescopes, they have been amazed at the seemingly infinite size of the universe. Of course we know that the universe is not infinite because it had a beginning according to Genesis chapter 1. But the incredible size of the universe has raised doubts about a young universe. "How could the universe be young when the nearest star to our planet is of such distance that it would normally take hundreds of years for light to travel from there to here?" This has been a major stumbling block for a lot of people. Let's address this.

1) First of all we do not actually know how long it takes for light to travel from there to here, because we have never measured it. Our perspective is from here to there. So we cannot be absolutely sure from a scientific point of view.

2) The creation event was not a natural one, but a unique, supernatural and historical event performed by God Himself. The universe did not come into existence naturally. According to Genesis 1:1 and 1:14-18, within a period of four literal days, God spoke the space into existence (See also Psalm 33:9), and then on day four he spoke the sun, moon and stars into existence. According to Genesis 1:17 God then placed them into the space for a specific purpose – to give light on the Earth. Again this did not take place naturally. Remember that God is not subject to the physical laws He created. This characteristic that belongs to God only is called "transcendence". All throughout history and the Scriptures God has done things that went beyond or counter to what we would expect in a natural world that exhibits physical laws. He caused the waters of the Red Sea to part. He caused the storm over the Sea of Galilee to stop. And He rose from the dead. These are all things that show that God Himself has power over the things He created. The stars were placed by God into the expanse of the space by a

supernatural act. They did not evolve into place. This is the big mistake that many make when asking the question about light, time and space.

3) One of the many things that bothered Einstein about the physical universe is the confirmed fact of an expanding universe. The galaxies are moving away from us. If we apply the uniformitarian principle of "the present is the key to the past", then eventually we are going to get back to the beginning. Einstein was extremely bothered by this. After all he was a scientist and scientists do not consider such things. Actually what bothered him was the idea that if the universe had a beginning, then Someone or something must have brought it into existence. For a uniformitarian physicist to admit that there just might be a powerful God who brought everything into existence is a very frightening thought! Why? Because that would mean that humans are accountable to a Creator and that naturalism cannot explain all things!

Activity F: Below is a table that summarizes the views of creation from both a Creationist perspective and a Naturalistic (uniformitarian; evolutionary) perspective. After reviewing this table, ask yourself, "Is there anyway that the record of Genesis can be harmonized with uniformitarian geology and evolution?

Order of Creation in Genesis vs. Order of Evolutionary Thought

Order of Creation in Genesis	Order of Evolutionary Thought
1:1 Time began when God created	Time is viewed as cyclical; no beginning and no ending; the universe in some form was always there
1:1 God Himself created; everything was brought into being from nothing by the word of God	The subject of God is irrelevant; everything unfolded naturalistically from other substances
1:1 The expanse or space (not the stars and heavenly bodies) and the Earth were created first	Matter and energy were present already and then other heavenly bodies slowly evolved beginning about 15 billion years ago and then the earth about 9 billion years after what geologists call the Big Bang
1:2 The Earth was developed first, but it was formless and empty	The Earth has taken about 4.6 billion years to develop, beginning with a molten ball at the beginning; all the necessary ingredients for formation and life were in the mix at the beginning
1:2 Darkness; there was no sun until Day 4 of Creation week	The sun (light) came first, before the Earth

1:2 Water was present at the beginning of the Earth's creation	Water came several billion years after the earth cooled
1:3-4 Light came into being by way of a direct word from God; God separated light and darkness; God was directly involved so our creation was a supernatural act	Darkness and light were there from the beginning of the Big Bang and is determined by celestial bodies like our sun; scientists decided in the 1800s that God was only involved in Earth's creation (Deism) so our Earth's development was a natural process occurring over billions of years
1:5 Light and darkness for the Earth was determined by 12 hour periods (evening and morning) and therefore one complete rotation of the Earth; verses 1-5 were accomplished in 1 literal day	Light and darkness are a product of celestial bodies like our sun
1:6-8 The Earth's oxygenated atmosphere was created on Day 2; water has oxygen and so oxygen was there from the beginning	The Earth's oxygenated atmosphere was developed over millions of years; in order for evolution to have occurred, the early atmosphere had to have been a reducing atmosphere (no oxygen) and so the "waters" were not oxygenated
1:9-13 The "seas" were gathered into one place and then the dry land appeared. The grasses, herbs and fruit trees sprouted with seeds on the Earth. The land vegetation was created on Day 3.	In evolutionary thinking the Earth is the focus. Plate tectonics of the land is what separated water into the oceans. Life began in the oceans. Grasses did not come on the scene for 5 $\frac{1}{2}$ billion years.
1:14-19 The stars, sun and moon were created on Day 4 in relation to the Earth. The purpose for these "lights" from the beginning was to separate day from night, mark the days, seasons and years. God decreed this and set this up.	In evolutionary thinking the sun and stars came first, then the Earth and then the moon. Seasons, days and years, were a natural part of the expanding universe. God had nothing to do with it.
1:20-23 Life in the oceans and birds in the skies were created on Day 5, after plants were created on Day 4. Life's diversity in the oceans and in the skies was there from the beginning of their appearance. Life was divided into kinds. This is why we are able to classify living things. Kinds have diversity within them.	Life began in the oceans, then amphibians and reptiles and birds evolved from dinosaurs, long after life first appeared in the oceans. Many land plants evolved after land animals and birds. Life is viewed as a continuous phylogenetic tree with everything interrelated and interconnected.

1:24-30 Land animals, including reptiles, amphibians and mammals were created on Day 6, after the plants on Day 4, ocean life and birds on Day 5. Man was created on Day 6. He existed at the same time as all the kinds of land-dwelling animals which would have included dinosaurs. Man was created in the image of God as male and female from the beginning and is therefore unique. God's authority over man was placed there at the beginning of his creation. Man's authority over creation was granted by God and so man is a steward of God's things.	The order of evolution: life in the oceans, amphibians, reptiles, birds, mammals and then man. Man came after the dinosaurs had died out. Many land creatures had existed for millions of years before man evolved. Man developed the idea of God and his beginning through evolutionary development. He is not to maintain uniqueness among the other animals, as he is one of them. Females evolved as a separate line (although scientists dare not say which came first until it is politically correct to say so). The idea of God evolved after man was on Earth for awhile. Man should not exercise authority over the Earth, but he usurps it. Everything is a connected system – nothing higher or more privileged than another.
1:31 The creation was finished by the end of Day 6. Everything was in perfect harmony with no death or struggle.	The "creation" is never finished. It is continuously evolving – a cyclical process. Death, decay and struggle are the vital processes of evolution, not of sin.

So, in brief, the order of creation vs. evolution goes like this:

God has always existed	God is irrelevant
Time ("In the beginning")	Time is cyclical; always existed
Space and earth & then light	Matter, space always there; earth subsequent
Water from the beginning	Water came sometime after
Seas and then dry land	Dry land and then seas
Plants, grasses and trees	Ocean life, then land animals and plants
Stars, sun and moon after the earth	Stars, sun before earth and then the moon
Ocean life after land plants	Ocean life first, then land plants
Birds on same day as ocean life	Birds after many land plants (came from dinosaurs)
Land animals after birds	Dinosaurs after land animals and then birds
Man on same day as all land animals	Man only after amphibians, reptiles, birds and mammals exited and evolved
Creation finished and perfect	Creation never finished and is always struggling
Creation completed in 6 days	Evolution still occurring after 15 billions years

Activity G: *Read the following Scriptures and record your answers in your notebook. Answers are in the Answer Key.*

1. *Genesis chapter 3*
a) *Describe in your own words the history of events that are recorded in this chapter.*
b) *Read Romans chapter 5 and 8 and describe how these two chapters relate to Genesis chapter 3?*

2. *Psalm 104:5-9*
a) *Describe the order of events*
b) *Which section describes the creation?*

3. *Hebrews chapter 1*
a) *Which sections tell us that a creator was involved in the origin of our planet? How was He involved?*
b) *Which sections identify this creator? Who is He?*
c) *Which section tells us what on-going part this creator plays in our planet? Does this view differ from the idea of Deism? If so, how does it differ?*

4. *What is the significance of the Bible? Why do we read it?*

5. *Exodus 20:1-11*
(a) *Who is speaking in this section?*
(b) *In light of this passage, what view would Moses have had of the creation of the world?*

Biblical Geology: Establishing a Proper Framework for Interpreting the Earth

The Foundation for any building must be solid. Jesus taught that for our lives, it was important to observe whether we were building on rock or sand. For Christians, that foundation is the Bible. But how do we apply it when studying the Earth? That is called the framework. It is the application of what the Scriptures way to what we observe around us.

Have you ever watched a house being built? The one thing that takes the most time and attention is the *framework*. Get this wrong, and the entire structure will not be sound. Because no human was there when the earth was created or to observe the rock strata being formed, we must have a proper framework in order to correctly interpret earth history.

As no eyewitnesses were around during Creation Week or when life supposedly evolved 550 million years ago, we have to construct a framework for interpreting what we observe now.

There are two views on the origin of life and man – Creationism and Evolutionism – we call them "isms" because they are beliefs about the past. The real questions are, which one is most reasonable? Which one best fits what we observe in nature?

Creationism	Evolutionism
Life has been complex since the beginning. It must have been created by an all-wise God	Life arose by chance and organized itself from simple to complex. No god was involved.

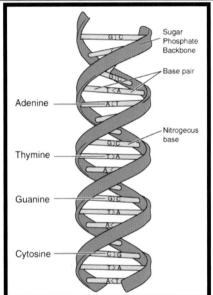

All of life has the complex structure on the left. This is called DNA. It holds information like a library, gives instruction like a computer to many different machines called proteins that build the various parts of the body that all coordinate together to perform life's functions. The chance that something this complex would develop by chance is less than **zero**. We call this "design". And it follows that a design must have a designer. This is Creationism.

Take Quiz #2, pg. 101

II The Origin and Nature of the Earth
Section 2: The Chemistry

Virtually everyone, scientist, atheist, and devout believer, agrees that the Earth we live on is a unique planet and a special place in our universe. As far as scientists have been able to discern, there is no other place like it – anywhere, period! Consider these facts:

- Earth is the only known planet to contain life – All of the things we need to survive are provided under a thin layer of atmosphere that separates us from the uninhabitable void of space.
- The temperature is just right for the maintenance of all life on Earth. If the overall temperature were to shift a few degrees either way, life as we know it would be impossible.
- The Earth is just the right distance from the sun (93,000,000 miles). If the Earth were to move any closer or further from the sun, it would either burn up or freeze.
- Although there are over a hundred known elements that make up our Earth, there are just 8 that form the majority of our Earth and they are in just the right proportions. If we had too much oxygen, life would not be possible. If there was more hydrogen than oxygen, life would not be possible.
- The Earth's atmosphere is a unique combination of layers that shield us from the sun's harmful radiation and from excessive heat and cold. If this delicate balance was to be altered, life would not be possible.

All the evidence points to a wonderful design and designer responsible for our magnificent planet we call Earth.

The study of the universe is called cosmology; the study of the cosmos; the well-ordered universe, as the Greeks defined it. The Greeks recognized the amazing design of the universe hundreds of years ago. The Greeks recognized something or someone more powerful than creation itself was behind the creation and maintenance of the universe. All ancient peoples recognized this, including the ancient Hebrews. Let's begin our study of the cosmos with discovering what the Earth is made of.

The Most Abundant Elements (atoms) in the Earth

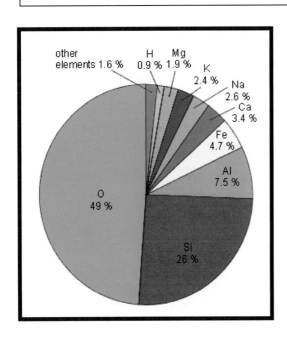

- Oxygen – 49%
- Silicon – 26%
- Aluminum – 7.5%
- Iron – 4.7%
- Calcium – 3.4%
- Sodium – 2.6%
- Potassium (K) – 2.4%
- Magnesium – 1.95
- Hydrogen - .9%
- Other elements – 1.6%

The **elements** (atoms) are the basic building blocks of matter. Read **Genesis chapter 1:1**. From this statement we see that right from the beginning space and earth were made of elements. For example the mineral quartz, an abundant mineral on earth is made of the elements oxygen and silicon. Water is made of hydrogen and oxygen. The major elements of creation are listed above. Notice that the major elements that form most of the Earth are absolutely vital and in just the right amounts for our planet.

The Atom

No one really knows what the atom looks like. The Greeks thought that there was a point which matter could no longer be broken down into parts. They called this the "atom", meaning, not able to be cut.

Matter, the Created Stuff of Genesis – "In the Beginning, God created..."
Matter is anything that occupies space. Matter, including all of life, is made up of atoms. An atom has three parts:

1. The **proton** – a positively charged particle in the nucleus of an atom
2. The **neutron** – a neutrally charged particle in the nucleus of an atom
3. The **electron** – a negatively charged particle outside of the nucleus of an atom

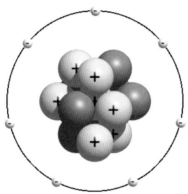

Atoms – are arranged in many precise patterns to form the different elements like oxygen, hydrogen and helium. To date there are 88 naturally occurring elements and 15 man-made elements.

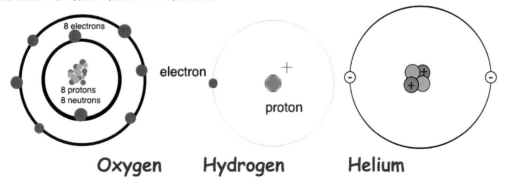

Oxygen   **Hydrogen**   **Helium**

One of the major questions in physics is, "What holds the atom together?" Have you ever wondered why things don't just fly apart?

*Activity H: Read **Hebrews 1:1-3**, and see if you can answer the above question using your Biblical framework.*

Man is also made up of atoms arranged in precise combinations and order. But he has one thing that no other living thing has – he was made in the **image of God**. That is something unique and does not consist of other created things or atoms. Something about man will live forever. No one has been able to really identify, from a scientific point of view, everything that makes up the image of God. This is beyond science.

Activity I: Read Genesis chapter 1:26. Describe some of the things that make man a unique creation. Write these down in your notebook.

Some more atoms:

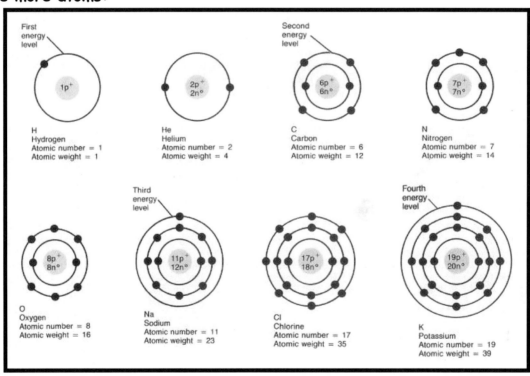

Minerals are combinations of elements (atoms). For example halite (rock salt) is made up of the element sodium (Na) and the element chlorine (Cl), arranged in a very special way. Sodium (a metal) by itself is very explosive and dangerous when put into water; chlorine is a poisonous gas by itself. But when combined in a special way, they give us an important ingredient in our food – salt. Jesus told us in the gospels that salt is good.

Activity J: *Do some research to try and find out why people today say that salt is a bad thing. Now, do some research to try and find out what is healthy and good about salt. Why is that?* *Answers in the Activity Answer Key.*

Halite or NaCl = sodium chloride (rock salt)

There are over 3,500 different minerals, but only about 12 that make up the rock-forming minerals. All other minerals that appear in rocks are incidental. That is, they show up sometimes and sometimes they don't. The rock-forming minerals are divided into two groups (light and dark) and these make all the rocks light and dark.

Study the chart below:

The Lighter Colored Minerals	The Darker Colored Minerals
Quartz Jasper (type of quartz) Potassium feldspar Muscovite mica Sodium feldspar Calcite (typically associated with sedimentary rocks)	Olivine Amphibole (Hornblende) Pyroxene (Augite) Biotite mica Calcium feldspar Magnetite (iron)

The Rock-forming Minerals

There are over 3,500 different minerals. But you only have to learn just twelve. Why only eleven? Even though there are many different minerals, about twelve make up most of the rocks that form our Earth. These are called the **Rock-forming minerals – quartz, jasper, potassium feldspar, sodium feldspar, calcium feldspar, biotite mica, muscovite mica, olivine, pyroxene, amphibole, magnetite (iron), and calcite.**

Activity K:

1. *Take out the bag of rock-forming minerals from your kit and study them carefully, using your magnifying glass. Describe what you see.*

2. *Sort the minerals according to light and dark. Which ones are light colored and which ones are dark colored?*

3. *Identify the elements that make up each of the rock-forming minerals and write those out. You may use your dictionary and/or encyclopedia and internet to help you. How many elements can you identify from the box entitled, "The Most Abundant Elements in the Earth"?*

4. *These elements make up the crust of the Earth and the things we eat. Get some food items from your kitchen and check out the nutrition labels. How many of the elements do you recognize from the box entitled, "The Most Abundant Elements in the Earth"?*

5. *Caution: Never put anything in your mouth to taste or investigate unless it is suitable for human consumption.*

The Rock-forming Minerals – the majority of rocks are made up of about 12 minerals called "rock-forming" minerals. These minerals are made up of the most common and abundant elements in the Earth:

1. Quartz (SiO_2) – made of the elements (Si) silicon and (O) oxygen
2. Jasper (SiO_2) – in addition to silicon and oxygen, it contains iron
3. Potassium feldspar $(KAlSi_3O_8)$ – made of (K) potassium, (Al) aluminum, (Si) silicon, and (O) oxygen
4. Sodium feldspar $(NaAlSi_3O_8)$ – made of (Na) sodium, (Al) aluminum, (Si) silicon, and (O) oxygen
5. Calcium feldspar $(CaAl_2Si_2O_8)$ – made of (Ca) calcium, (Al) aluminum, (Si) silicon, and (O) oxygen
6. Biotite mica $(K(Mg,Fe)_3(AlSi_3O_{10})(F,OH)_2)$ – made of (K) potassium, (Mg) magnesium, (Fe) iron, (Al) aluminum, (Si) silicon, (F) fluorine, (O) oxygen and (H) hydrogen (the magnesium and iron give it its black color)
7. Muscovite mica $(KAl_2(AlSi_3O_{10})(F,OH)_2)$ – made of the exact minerals as biotite mica with the exception of magnesium and iron
8. Olivine $((Mg, Fe)_2SiO_4)$ – made of (Mg) magnesium, (Fe) iron, (Si) silicon, and (O) oxygen
9. Pyroxene $((Ca,Na)(Mg,Fe,Al,Ti)(Si,Al)_2O_6)$ – made of (Ca) calcium, (Na) sodium, (Mg) magnesium, (Fe) iron, (Al) aluminum, (Ti) titanium, (Si) silicon, and (O) oxygen
10. Amphibole $((Ca,Na)_2(Mg,Fe,Al)_5(Al,Si)_8O_{22}(OH,F)_2$ – made of (Ca) calcium, (Na) sodium, (Mg) magnesium, (Fe) iron, (Al) aluminum, (Si) silicon, (O) oxygen, (H) hydrogen, and (F) fluorine
11. Magnetite (iron) (Fe_3O_4) – made of (Fe) iron and (O) oxygen
12. Calcite $(CaCO_3)$ – made of (Ca) calcium, (C) carbon, (O) oxygen (primarily in sedimentary rocks)

The rock-forming minerals are part of broader groups of minerals called, "Mineral Families". All of the rock-forming minerals above belong to the **Silicate Family** of minerals because of the shared characteristics of (Si) silicon and (O) oxygen, except calcite. It belongs to the Carbonate Family (CO_3) of minerals. Did you notice just how abundant the basic elements are in these minerals?

Minerals are grouped according to families because they share the same characteristics. Is this by chance or by design? What do you think?

The Mineral Families (pictured below from left to right):

The Silicate Family – typical mineral is quartz; all contain silicon & oxygen

The Carbonate Family – typical mineral is calcite; all contain carbon & oxygen

The Phosphate Family – typical mineral is apatite; all contain phosphate & oxygen

The Halide Family – typical mineral is halite; all contain at least one of the following: fluorine, bromine, iodine, chlorine

The Sulfide Family – typical mineral is iron pyrite (iron sulfide); all contain the element sulfur as a sulfide (S_2)

The Sulfate Family – typical mineral is gypsum (calcium sulfate); all contain sulfur and oxygen (SO_4)

The Oxide Family – typical mineral is hematite or iron ore (iron oxide); all contain oxygen

The Borate Family – typical mineral is borax; all contain borate and oxygen (BO_3)

The Native Element Family – typical mineral is gold (mainly one element occurring naturally)

Take Quiz #3, Pg. 101.

The Mediterranean Sea

II The Origin and Nature of the Earth
Section 3: Water, Water Everywhere

Water

Many geologists will teach that water developed on earth over billions of years as substances from volcanoes and magma slowly changed our atmosphere. But if this is so, then why haven't other planets developed water? After all, according to the Big Bang idea in modern geology, millions of other planets have been around a lot longer than Earth has. But as far as we know, water in the abundance found on Earth, is only found here! Water is a unique and absolutely critical part of our existence. Is it therefore valid to consider that perhaps it was brought into existence by an all-knowing and all-powerful God? Isn't this a part of good detective work to consider other ideas besides uniformitarian ones?

Activity L: Read Genesis 1:1-2. This Scripture states that at the beginning of creation, the earth was most likely a watery sphere. Water is a very special substance that God created and it can do a lot of different things. What might this say about water's role in the on-going maintenance of the Earth and its systems? Be sure to record your answer in your notebook. Answers in Answer key.

No other planet has water like Earth. Water is made of elements or atoms. Water has 2 atoms of hydrogen and 1 atom of oxygen. Its unique structure allows it to hold different things in balance suitable for life on earth. Below is a picture chemists use to illustrate the idea of a molecule of water (H_2O):

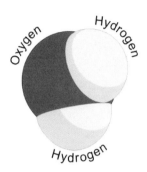

A water molecule has 2 atoms of hydrogen and 1 atom of oxygen combined in a special way to produce water. *Is this special design a product of naturalistic chance or of an act of the God as taught in Genesis?*

Read 2 Peter 3:3-9.

Peter, in this passage of Scripture, tells us that the Earth was created out of water and by water. The two pictures below give some kind of an idea of what the Earth might have originally looked like from a Biblical geology framework. From this passage we also see that water was an integral part of the Earth from the beginning. The dry land "appeared" on Day 3 of Creation week. So it was already present when Earth was created, but covered by water.

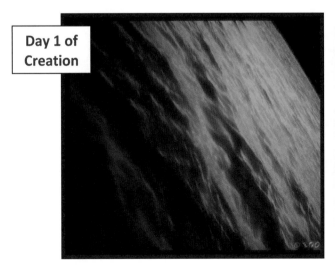

Day 1 of Creation

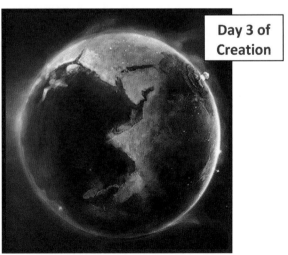

Day 3 of Creation

Acids, Bases and Water

Many substances are either an acid or a base in different proportions. An acid has a hydrogen (H) atom in its structure while a base has an atom of hydrogen and an atom of oxygen grouped together (OH). Acids taste sour while bases tend to taste bitter. These two things are an important part of our life existence.

Activity M:

Vinegar is an acid. Place some household vinegar into a spoon and put some on your finger. Now taste it. What do you notice? Now place some baking soda (a base) on your finger and taste it. What do you notice? **Caution: while many substances are weak and harmless, many things are not, such as hydrochloric acid. These can be extremely dangerous and should only be handled according to directions. Please be very careful when experimenting with different substances. Always read labels. Never directly taste or smell anything that is not meant for that purpose. Here is a personal story: When I was much younger, I was experimenting with a chemical called ammonium hydroxide – a very powerful base. Not paying attention, I held the bottle to my nose and proceeded to smell it. It was so powerful that it temporarily took away my ability to breathe. I could not catch my breath. I thought I was going to die. Fortunately in a short time my ability to breathe returned. This was an unfortunate lesson but an extremely important one! Please learn from it.**

The pH Scale

Chemists have developed a scale to measure the acidity or base of a substance. The scale is constructed on the basis of 1 to 14 with 1-6 being acidic, **7 (water) neutral** and 8-14 being basic (base). This scale measures the degree of acidity or base in a given substance. Example: The substance HCl (1 on the pH scale) has one atom of hydrogen and one atom of chlorine. This is hydrochloric acid, a very strong acid. Baking soda ($NaHCO_3$) is a base. It measures 9 on the pH scale. Each of these substances have unique purposes in God's creation. But here is the amazing thing. Now, here is the wonderful thing about water. Water has the ability to neutralize or soften both acids and bases. This ability speaks loudly of design and protection of an all-knowing Creator. To imagine this coming about by chance requires a blind faith. Water is a product of design, not naturalistic chance. Water is truly an amazing thing.

Did you know?

- The earth is the only place in the universe known to have liquid water
- The earth is 70% covered by water
- Only about 1% of the world's water is ready to drink (but this is enough). About 97% is too salty and 2% is ice

- Pure water is colorless, odorless and tasteless. Tap water may contain small amounts of salts and gases, which give it a taste
- Australia is the world's driest continent
- Only 1% of household water is used for drinking in Western countries. The rest is used in bathrooms, and on the garden, etc
- A house toilet flushes about 39½ gallons of water per day
- A tap dripping one drop per second wastes about 8 gallons of water per day

The pH scale

Acids   **Neutral**   **Bases**

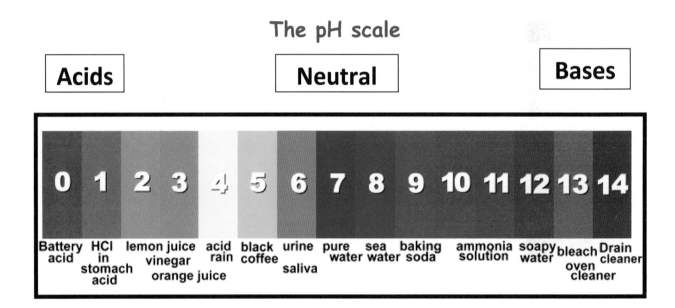

The pH scale tells us how much acid or base something has. The two most dangerous items on the pH scale are battery acid and drain cleaner. Both are dangerous by themselves. Both acids and bases are caustic. Causticity is the property of a substance that causes corrosion. Corrosion is the destruction of materials. **Water** has the amazing property of being able to neutralize both acids and bases. It is vital to life. Without water, no life would be possible. Acids and bases are necessary for life too.

Activity N: Write out examples of 5 acids and 5 bases. Explain how each of these helps our lives, but also if consumed in the wrong way, hurts our lives. Explain how water helps keep acids and bases from becoming harmful.

Lemons are acidic. They taste sour.

Take Quiz #4, pg. 102.

Our atmosphere, as viewed from the Space Shuttle.

II The Origin and Nature of the Earth
Section 4: The Magnificent Atmosphere

Earth's Atmosphere
Read Genesis 1:6-8.

The air is made up of mainly nitrogen, some oxygen, and a tiny amount of other gases such as carbon dioxide. These are in just the right amounts. If there were a lot more oxygen, then a single spark could set the world on fire. If there were a lot more nitrogen, we would suffocate. Carbon dioxide is essential for plants to live, but is deadly to humans in large quantities. Earth's atmosphere has only a tiny amount of carbon dioxide, which is plenty for all the plants, because they use it very efficiently, and this small amount is harmless for us. By comparison, the atmosphere on Mars is 95% carbon dioxide. This is one reason why people could not live on Mars without special breathing equipment. The atmosphere also contains water vapour which condenses to form clouds which give us needed rain. Our atmosphere does something else – it shields us from harmful radiation from the sun and space. It protects us from burning up because of the sun's intensity. And, the atmosphere provides the air we breathe. How could such a thing develop naturally? It seems that our Earth is built just right for the right thing in the right place with all the right ingredients just right for life. This is design, not naturalism.

Activity O: Study the following diagram. Notice the different layers of our atmosphere. Such coordination of these layers could only mean one thing – that God specially designed our atmosphere for our survival.

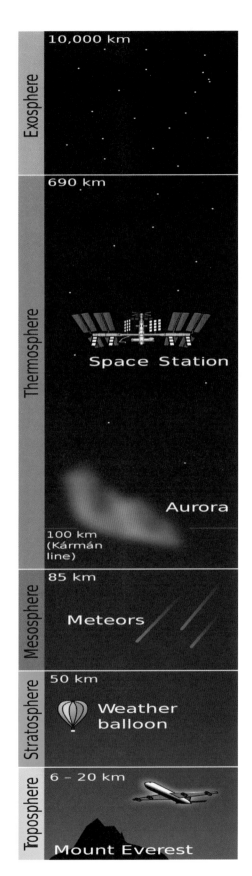

Considered to be outer space; here the density is so low that heavier particles escape very easily. Without this layer, life would be in constant danger of being bombarded by every object or particle in outer space; acts as kind of a shield

Up to about 620 miles above sea level; here we are protected from ultraviolet radiation – without this layer, life would not be possible

Up to about 56 miles above sea level; here the air temperature becomes colder the higher up – just the opposite of the stratosphere

Up to about 30 miles above sea level; here the air temperature becomes warmer the higher up – just the opposite of the troposphere

30,000-60,000 feet above sea level; *contains about 99% of the water vapor necessary for life;* air temperature becomes cooler the higher up. Water vapor (H_2O) contains oxygen

Earth's magnetic field – amazing and protective

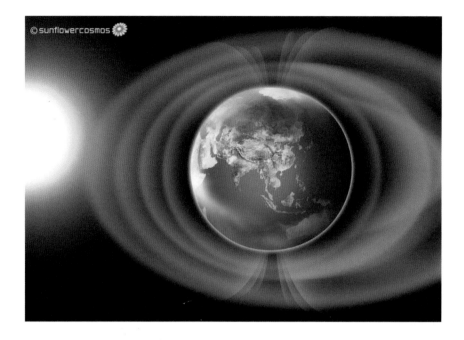

There are just a few things that you should know about Earth's magnetic field.

1) It seems to be a very special force that surrounds our globe like a huge magnet
2) It is invisible but plays a vital role in preserving our planet
3) Along with the atmosphere, it helps shields us from harmful radiation coming from the sun and stars
4) The Earth's magnetic field protects us from solar wind. Without this field, solar wind would strip away our ozone layer and make life impossible
5) The Earth's magnetic field is decaying! Studies have shown that the Earth cannot be billions of years old because we know that this magnetic field is decaying at too fast of a rate. Otherwise we would have ceased to exist many, many years ago. The magnetic field would have all decayed long ago
6) The presence of a protective magnetic field around our Earth is a part of God's design, not a product of chance naturalistic processes

The Interior of the Earth

Do we really know what the interior of the Earth looks like? No one has ever been there, despite the story written by Jules Verne entitled <u>Journey to the Center of the Earth</u>. The deepest drilled hole, located in Russia, is the Kola Superdeep Borehole. It is 7.6 miles deep and it took 20 years to do it! And it is still a long way through the rest of the crust to the mantle. However, through some mathematical computations and use of seismic waves, scientists have been able to construct an idea of what the interior of the Earth might look like.

The density of rocks and minerals plays an important part in coming up with an idea of what the interior of the Earth might look like. Remember our 12 rock-forming minerals? Some of the densest minerals that make up rocks are magnetite (containing the element iron) and olivine. Iron and olivine form the darker rocks such as gabbro and basalt. Olivine itself is made of the elements iron and magnesium. Comparing the density of these minerals and the rocks that contain them with mathematical computations and seismic waves, the density of the interior of the Earth seems to be very close to these minerals. So, here is an illustration of what the interior of the Earth might look like.

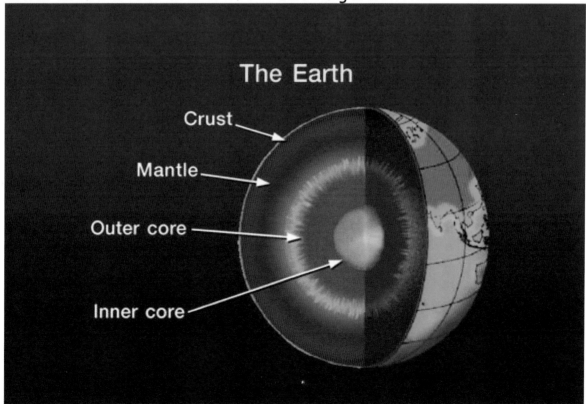

1. The **crust** of the Earth is very thin in relation to the overall size of the Earth. It is estimated to range to about 37 miles thick. This is called the lithosphere (Greek, *litho*, stone, and *sphaira*, sphere). The crust consists of two parts, the continental crust, the visible rocks all around us, primarily of granite (made of the minerals quartz, feldspar and mica), and the oceanic crust made up of mostly basalt lava (made of iron and olivine). These two parts of the crust of the Earth consist of the rocks that we can readily collect and observe.

2. The **mantle** is thought to run to about 1,796 miles deep. Its density seems to be similar to the density of the mineral olivine. But scientists have noticed an abrupt change at about 1,796 miles based on seismic waves.

3. The **core** of the Earth - The density increases to about the consistency of iron and nickel. And this seems to take up the rest of the 2,163 miles to the center of the Earth.

Of course we do not know these things for sure, but they seem to fit what we have been able to observe and calculate to date. One thing is for certain, the Earth has been designed as a very special place for man. In addition to what we learned about the atmosphere, the core of the Earth being made of the element iron would make perfect sense and would explain Earth's wonderful magnetic field. It invisibly protects us from some of the high energy particles from the sun (solar wind) from overheating the planet; it allows water to exist as gas and as a liquid; and it helps both us and animals in navigation.

Activity P:
1. *Take out the olivine and magnetite, found in your kit, and study them with the pocket microscope or some other magnifying glass. Write down your observations in your lab book.*
2. *Take out the granite and basalt, found in your kit, and study them with the pocket microscope and write down your observations in your lab book.*

Take Quiz #5, pg. 102.

II The Origin and Nature of the Earth
Section 5: Building Blocks – The Rocks

Rocks

Now we come to my favorite part of geology (Earth science) – Rocks! Rocks make up the crust of the Earth. There are two parts of the Earth that we are able to explore and map. The Earth is considered to have a mantle and a core, but these have not been directly observed. We have only actually observed parts of the crust. So for now, we will only talk about the crust. There are two types of crust:

1) **The continental crust** consisting mostly of a plutonic rock called granite. Most of the continental crust is covered by trees, grasses and sedimentary rocks. But here and there large areas of granite have pushed up high into the air. One of these huge granite formations is found in Yosemite National Park in California.

Activity Q: Here the Yosemite National Park Geology Kit from **Northwest Treasures (Appendix F)** would be a great kit to work through. Or, research Yosemite National Park and write a short report on the formations there.

2) **The oceanic crust** consisting mostly of a type of lava called basalt.

Activity R: Obtain a map of the ocean floor.
   1. What kinds of things do you notice?
   2. How would you explain the valleys/canyons on the ocean floor from a Biblical Geology framework? *Answers in the Answer Key.*

Rocks are divided by geologists into three categories: **igneous rocks, metamorphic rocks** and **sedimentary rocks**. I have divided them a little differently because there is confusion as to their origin and formation. Let me explain:

Igneous Rocks

The word igneous comes from the Latin word for fire. Uniformitarian geologists believe that some rocks originated when they think the Earth was a molten ball of magma and fire billions of years ago. They call these rocks plutonic or intrusive igneous rocks. Slowly cooling over millions of years, uniformitarian geologists think the magma became igneous plutonic rocks – the stuff of the Earth's crust. Obviously if this is true, then the Scriptures cannot be true and its explanation for the origin of the Earth is false! But now here is a great question to ask. *"Have geologists ever witnessed these kinds of rocks being formed this way?"* The answer to that is a resounding "NO". Then why do geologists teach this? Because their uniformitarian worldview or framework influences them. Because the Genesis account of creation has been rejected from consideration, modern geologists are left with a naturalistic uniformitarian view for the origin of the Earth.

So, how would a Biblical framework explain the origin of these rocks? Very simply. **Read Genesis chapter 1.** Remember that the elements, the basic building blocks of matter, were created during the 6 days of Creation. Elements form everything we see. This includes the rocks. So, at least some rocks were formed during this week of Creation. I say some, because other rocks may actually have been formed during the Genesis Flood which we will discuss later.

I divide igneous rocks into *two* different groups: **Plutonic** (igneous) rocks and **volcanic** (igneous) rocks. But only one of these rocks have geologists seen forming. Can you guess which one? That's right, **volcanic** rocks. I like the word plutonic, after the Roman god of the underworld, because it describes what I think Genesis teaches. Rocks that were part of the original foundation of the Earth are found deep underground; plutonic rocks like granite and gabbro. Then at some point some of these rocks were pushed up and became part of the Earth's surface. I say "some", because we don't exactly know which rocks was part of the original creation and which rocks were formed during the Genesis Flood. For purposes of organization I have color-coded the rock types below so that you can keep them straight and organized:

Sedimentary (brown)
Plutonic (dark blue)
Volcanic (orange)
Metamorphic (red)

Plutonic Rocks. Let's take a look at the plutonic rocks. Plutonic rocks are identified by their grain or mineral size. Rocks in which the minerals are easily recognized by their minerals are called plutonic.

1. **Gabbro.** Funny name isn't it? This rock was actually named for a town in Italy of the same name where gabbro typifies this particular rock. Gabbros are generally dark in color because of the dark rock-forming minerals it contains. Gabbros are coarse grained where the minerals are easily seen with the naked eye.

2. **Granite.** Italian granito, from past participle of granire to granulate, from grano grain, from Latin granum. To make a long story short, granite is a coarse grained rock where the minerals are easily seen with the naked eye. Granite consists of the light colored rock-forming minerals.

Gabbro and Granite

Activity S:
Take out the plutonic rocks in your kit and a magnifying glass and see how many minerals you can see. Don't worry about whether you know what they are, but whether they are easily seen with the naked eye. Record your answers.

Activity T:
Take out the bag of rock-forming minerals from your kit and see if you can match the minerals with what you see in each of the plutonic rocks. Record your answers.

We will discuss the other rock types (volcanic, metamorphic and sedimentary) when we discuss the Genesis Flood.

Activity U:
Read Genesis chapters 1-2. Write down as many things as you can that tell us how different the Earth was from today. And then, answer the following questions. Answers in the Answer key.
1. *Is death or sickness mentioned anywhere? If so, under what circumstances are they included?*
2. *Is the word sin mentioned anywhere? Is war or disease mentioned anywhere? Is the word corruption used anywhere in Genesis chapters 1 and 2?*
3. *What might these things have to do with changing the Earth in the past from what it is today?*

Activity V:
Read Genesis chapter 3. Now read Romans chapters 5 and 8. How are these three chapters related? What might these three chapters have to do with geology? Record your answers. Answers in the Answer Key.

The Origin of the Earth and its various systems is summarized in Genesis Chapter 1:

- Time and matter had a beginning
- God was already there; He is eternal; He had no beginning
- God was involved from the very beginning of Earth history and continues to be involved
- Days were complete, 24 hour revolutions of the Earth with evening and morning; furthermore, the "days" are ordinal. This means that something was completed during that specific period of time and it was "good".
- There were 6 of these day-cycles and on the 7th, God rested; things were complete; finished before any death or corruption
- Space and earth came before stars, planets, sun and moon
- Water and soil were a part of the beginning of the creation of the Earth
- Life is divided into kinds which reproduce after their kinds; variation is observed, but only within kinds
- Man was created as a separate kind, not related to any other kind. What gives him his uniqueness is the fact that he has been made in the image of God
- Although we don't know for sure, most of the plutonic rocks were probably formed during this week as part of the foundation rocks of the Earth

Overall Review of Genesis Chapter 1 - The Beginning of Earth History

Activity W: *Summarize Genesis Chapter 1 by briefly listing what was created on each day. Record your answers. Answers in the Answer Key.*

1. *Day 1*
2. *Day 2*
3. *Day 3*
4. *Day 4*
5. *Day 5*
6. *Day 6*
7. *How do we know that the day in Genesis chapter 1 was a 24 hour normal day?*
8. *Read Exodus 20:1-10. What did God directly speak to Moses concerning the work week that would make one view the day as a 24 hour normal day?*
9. *Read Genesis 1:1 and 1:14-19. Which came first, the galaxies or the earth? How does this conflict with the Big Bang worldview? How did the stars, sun and moon get into space?*
10. *Read Genesis Chapter 1:11-12, 20-22, 24-25, and 26-28. What are the major divisions of living things that God created?*
11. *Where would dinosaurs fit into these divisions?*

Day 1: Earth, space, time, light; **Day 2**: Atmosphere; **Day 3**: Dry land and plants; **Day 4**: Sun, moon and stars; **Day 5**: Sea creatures and flying animals; **Day 6**: Land animals and man

Take Quiz #6, pg. 102.

The Genesis Flood as presented by the Scriptures would have included massive destructive geological forces the world had never and will never witness again. The entire surface of the Earth would have been totally rearranged beyond recognition from what it had been.

III The Genesis Flood – The Historical Turning Point of Earth History
Section 1: Biblical Integrity

Vocabulary:

Look up and write out the meanings of the words below:
Catastrophic (Catastrophe)   Tectonic   Global   Metamorphic
(Metamorphosis)   Strata   Deposition   Fault   Rift
Inundate (Inundatory)   Receed (Receeding)   Extrude (Extruded)
Lava   Viscous (Viscosity)   Magma   Phenocryst   Orogeny
Guyot   Fold (in geology)   Aquifer

One of the greatest challenges for the believer today is to reconcile what the Bible teaches about a special creation about 6,000 years ago and a global flood that killed all plants, land-dwelling animals and man about 4,500 years ago with what modern geology states: The Earth was formed about 4.6 billion years ago and there was no global flood. In other words it is very old and its formations were formed over hundreds of millions of years by gradual repetitive geologic processes, not by a global flood! This opinion is shared by almost all scientists today. What was once believed by Isaac Newton and other great scientists, that the Earth is young, has not only been rejected but is also mocked in many scientific circles today. We have to resolve this issue because the very integrity of the Bible is at stake.

What the Bible Records about a Global Flood

Activity X: Read Genesis chapters 7-8 and record your answers to the questions below. Answers in the Answer Key.

1. *What statements from these chapters do you observe that teach that the Flood was more than a simply local flood?*

2. *How was Earth affected by the Genesis Flood?*

3. *What statements tell us that the Flood was geological?*

4. *Look at the historical record of the early Biblical patriarchs. How many years from Adam to Noah were there? How many years ago did the Flood take place?*

Green River Gorge in western Canyonlands National Park, part of a 130,000 square mile sedimentary landform. How would a Biblical Geology framework explain this vast formation? *Photo courtesy of the USGS.*

Take Quiz #7, pg. 102.

This is an artistic rendition of what it might have looked like
when the fountains of the Great Deep opened.

III The Genesis Flood – The Historical Turning Point of Earth History
Section 2: The Flood – Just What Happened, and When?

The Stages of the Genesis Flood

Many people think that the Genesis Flood was only a localized flood that took place somewhere in the Middle East and only lasted for 40 days. But the Scriptures tell us that it was a global flood lasting over a year with devastating geological consequences! If one reads chapters 7-8 of Genesis, it will be readily apparent that the Flood can be divided into 2 stages. Each stage of the Flood brought on tremendous geological upheaval and change to our planet.

1. **Inundatory or The Flooding Stage** – this was the beginning stage and a very violent event. For 150 days the water shot up from below the Earth, pouring warmer water into the ocean of the past and rising to cover what would have been the mountains of that time. The Scriptures record that the water covered what would have been the highest mountains at that time by 22 ½ feet. This would have been the maximum depth of the Flood. No land animal, bird or air-breathing creature survived this devastating geological event, except Noah, his family and all the animals which were on the ark.

This stage of the Flood would have been responsible for tremendous amounts of sediments being torn up, transported and then deposited into layer after layer all over Earth. An example of this stage of the Flood would be the many layers of sedimentary rock in The Grand Canyon before it was a canyon. Where do you think all the water came from? Deep underground, all over the world are geological structures called aquifers. This word comes from the Latin for water, "aqua". Even under places that seem to be totally devoid of water are aquifers. These aquifers might have served two purposes in the days of the Flood:

a) A source of water (besides the ocean that was present at the time), and
b) One of the drainage places for the water (besides the present ocean basins) *Study the maps below and notice that abundant aquifers exist even under places like Africa and Australia.*

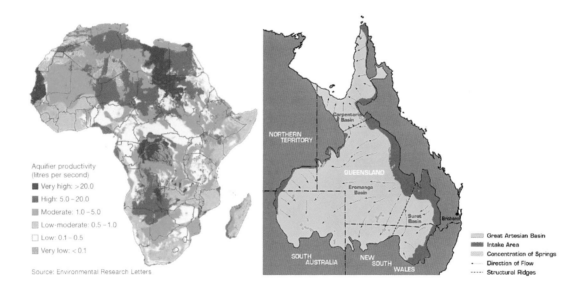

2. The Receding or Retreating Stage.

Activity Y: Read Psalm 104:5-9 and record your observations in your notebook.

Toward the end of the flooding or inundatory stage of the Flood, Psalm 104 states that at God's rebuke the mountains rose and the valleys sank down. This effectively describes:

a) *Mountain building or orogeny*
b) *Creating the ocean basins*
c) *More sediments being torn up, transported and redeposited*
d) *Canyon cutting*

Secular geologists tell us that most of our most famous mountain ranges like the Rocky Mountain Range and the Himalayan Mountain Range were formed recently in geologic time. Actually this is exactly what the Biblical framework would describe. Mountain ranges were formed (according to Psalm 104:5-9) in the end stage of the Flood. The building of these mountain ranges would have served two purposes:

a) To facilitate the draining of the flood waters that had been covering the Earth, and

b) To provide a boundary so that the waters would not cover the Earth again

Once the sediments of The Grand Canyon were laid down across a broad area, then the cutting action of the receding or retreating stage of the Flood would have gone to work, removing a large amount of sediments, still not quite hard enough to resist. The weaker sediments would have been cut away while the stronger sediments would have remained. Effectively huge "scars" would have been left in many places around the world. Besides The Grand Canyon, Monument Valley would have been carved at this same time, leaving the stronger or tougher rock in the form of isolated sandstone structures or "monuments".

The Grand Canyon and Monument Valley

Take Quiz #8, pg. 102.

This sandstone formation called "The Wave" is in northern Arizona. Notice the flat contacts between these successive formations of sandstone.

III The Genesis Flood – The Historical Turning Point of Earth History
Section 3: Unusual Land Forms

Other landform features of the Earth explained by the Genesis Flood

Folded sedimentary rock

Activity Z:

1. Get several pieces of bread. Next, spread peanut butter on them and put them together. We will use these to represent sedimentary layers that are still moist from the deposition of the Genesis Flood. Now very carefully and slowly push the edges of the bread and peanut butter together and observe what happens. Record what you observe.

2. Take several popsicle sticks and paint them different colors. Glue them together and let them harden. After the glue has dried and hardened, try bending the sticks. Record what happens. Answers in Answer Key.

Folded sedimentary rocks are all over the world and they are evidence of a global catastrophic flood that was responsible for the layering and the bending of the layers before they hardened.

Devil's Tower

I am sure you recognize this geological landform. It's Devil's Tower in Northern Wyoming. Secular geologists tell us that this is a volcanic neck (the inside of a volcano) and that it has been preserved after all the outer sediments and volcanic features have eroded away over millions of years. Let's sort out the science from the philosophy. Chemically the rock is hard, very hard basaltic type of volcanic rock. Basalt lava typically forms geometric columns as it cools and hardens. And it appears to be the remnant of a volcanic neck. How did it form? As no one witnessed its formation, we must start with a framework. There are several things to notice about this structure:

- The top of this volcanic remnant is flat
- The lava columns show very little erosive wear
- All of the sediment around this volcanic feature has been removed

The receding Flood waters do provide a great explanation for the formation of this remnant. The force of receding Flood waters would have sheared off the top of the volcano and removed the softer sediments around it. Because this was a fairly rapid and recent process we would expect the lava columns to look somewhat fresh with very little erosive wear.

Planation surfaces

Planation surfaces are some of the greatest mysteries of modern geology. What kind of slow geological process would erode the sides and valleys and leave a flat top? Why would the entire structure not show the same erosion patterns? Again using our Biblical framework of the Genesis Flood, the receding Flood waters would have planed sedimentary pliable layers built up from the deposition of the Flood, transporting the softer sediments and leaving the more resistant sediments behind. The same sort of shearing effect would have occurred as had happened to Devil's Tower.

Guyots (guy-oats) or Seamounts

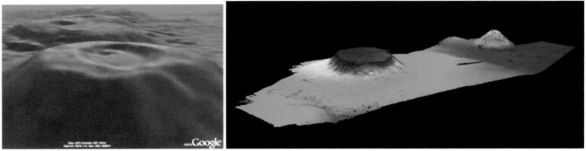

USGS and Google Earth; Computer generated image of a guyot or seamount. Oceanexplorer.noaa.gov

These are essentially flat-topped volcanoes. And typically they are found on the ocean floor.

Diagram of a Guyot (pronounced "guy-oat"), a flat-topped volcano, covered by water.

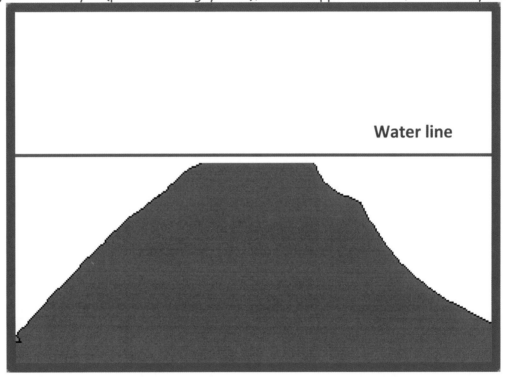

Secular geologists readily acknowledge the existence of guyots. The real question is what force could have sheared a volcano flat?

Activity AA: Using the Biblical Geology framework discussed above, describe how these landforms might have been formed? Record your explanation in your notebook. Answers in the Answer key.

Take Quiz #9, pg. 102.

The Painted Desert, Arizona

III The Genesis Flood – The Historical Turning Point of Earth History
Section 4: The Rocks of the Flood

The Rocks of the Genesis Flood

Volcanic Rocks.   Volcanic rocks are also identified by their grain or mineral size. Rocks where the minerals cannot be discerned with the naked eye are called volcanic rocks, after the Roman god of fire, Vulcan. Geologists have seen many of these rocks form and these rocks are associated with volcanoes or lava flows. These are the true igneous rocks, because we know that they are formed from very hot lava that has extruded from volcanoes or rifts in the Earth and cooled. Volcanic rocks are referred to as fine grained rocks. A person can see a general color to the volcanic rock, but usually cannot make out the individual minerals in the rock. One other type of volcanic rock can be a little confusing because some of its minerals can be seen.   Geologists call this type of volcanic rock, porphyry or porphyritic. The word means, "purple". Huh? It's a little confusing isn't it? I have included a volcanic porpyritic rock in your kit. Volcanic porphyritic rocks are quite common.

Activity BB:
Take out the **volcanic** *rocks in your kit and a magnifying glass and see if you can see any of the minerals that make up the rock. Record what you see.*

We call the mix of minerals in the volcanic rock the "groundmass". It is the overall rock that consists of the minerals that form the rock. Many volcanic rocks have incidental mineral crystals showing. These are called "phenocrysts". These are not the actual groundmass of the volcanic rock, but simply individual minerals called porphyry. These pheocrysts also help identify what type of volcanic rocks they are. Certain volcanic rocks have predictable phenocrysts.

1. **Basalt.** This word actually means "hard". Basalt is a very hard rock. Like gabbro, it is formed of dark minerals. In fact basalt may even be the lava equivalent of melted and cooled gabbro.
2. **Rhyolite.** This word means "flow". Rhyolite is a type of lava that is so viscous that as it cools and hardens it leaves flow patterns. It is the equivalent of granite. In fact rhyolite may even be the lava equivalent of melted and cooled granite.

Basalt **and** Rhyolite

Metamorphic Rocks

The word "metamorphic" is from a Greek work meaning change. It is used in the New Testament in Romans chapter 12: 2, "... be *transformed*...."

Metamorphic rocks are identified by the terms "foliated", meaning that there is visible banding or layering, and "nonfoliated" meaning that there is no visible banding or layering. In foliated rocks the minerals appear to have been rearranged into bands. Nonfoliated metamorphic rocks rather display a crystalline appearance. Many metamorphic rocks display reworking of the minerals such that they appear to have been deformed or stretched. They also appear to be rocks that have changed from some previous rock. For example take a look at the pictures of granite (a plutonic rock) and gneiss (a metamorphic rock) below and see why geologists might conclude that gneiss is really reworked or "recooked" granite.

Granite (plutonic rock) and Granite Gneiss (foliated metamorphic rock)

Limestone (sedimentary rock) and Marble (nonfoliated metamorphic rock)

What on Earth would cause rocks to change like this? Although metamorphic rocks have not been observed to be forming, geologists guess that these rocks have formed over millions of years under a great deal of heat and pressure. Biblical geologists would agree that it probably took a great deal of heat and pressure to produce these rocks. The big difference is the time frame. The Biblical chronology will not allow millions of years. So, what in Biblical history would have produced such intense heat and pressure to deform and change rocks? Well, of course, the Genesis Flood. Take a look at the drawing below and imagine the very first day of the Genesis Flood. The Bible says that all the fountains of the great deep burst open. (Genesis 7:11) This description implies that the Earth of that time went through catastrophic upheaval. In geology we call this tectonics. The Earth would have cracked and shook to its very core. Such Earth movement would have generated a great deal of heat and pressure as huge amounts of existing rock moved against each other. The Genesis Flood implies that there was rapid tectonics and thus rapid metamorphism. The next pictures are of the MidAtlantic Rift. The one below right runs through Iceland. Ever since the 1970s, mapping the ocean floor has revealed some amazing evidence for the Genesis Flood. Take a look at the picture of the ocean floor and notice the seams all around the globe. These are most likely the remnants of the breaking up of the fountains of the great deep.

 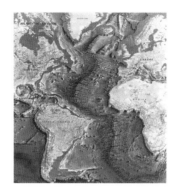 

Sedimentary Rocks

As the name implies, these rocks are made up of sediments that have been cemented together and then hardened into rock. Does it take a long time to turn sediments into rock? Let's investigate. Try the following activity.

Activity CC:
1. *Put four cups of sand in a bucket. Add four cups of water. What happened to the water?*
2. *Next, get a bucket of dirt. Make a small hole in the middle. Then, get a small bowl of wet sand out of your first bucket. Mix Epsom salt with the sand. Dump the mixture in the hole in the bucket of dirt and gently pat it. Wait two days and examine the mixture. Answer in Answer Key.*

Sedimentary rocks are rocks laid down by water and mud. 70% of the Earth's surface is covered with sedimentary rock. That's a lot of rock. Sedimentary rocks are grouped according to the terms **clastic**, **chemical** and **biochemical**.

Clastic Sedimentary Rocks

The word "clast" is from the Greek word for "broken". So, sedimentary rocks consist of other broken bits and pieces of rocks and minerals cemented together into a rock. The clastic sedimentary rocks are grouped according to the size of the clasts.

From left to right: 1. **Shale**; 80% of the sedimentary rock on Earth is shale. Shale is also called claystone or mudstone because it is made of *very fine particles of clay and mud.* It is very fine grained and extremely brittle. Shale may contain fossils. 2. **Siltstone**; consists of clasts that are *a little larger than clay particles.* 3. **Sandstone** has clasts *easily seen with the naked eye.* 4. **Arkose** is *coarse sandstone (more coarse than regular sandstone)* with the mineral feldspar as one of the cementing agents. 5. **Conglomerate** can consist *of rather large **rounded***

clasts called pebbles, cobbles and boulders. 6. **Breccia** is a *type of conglomerate, but with angular clasts.* These two clastic sedimentary rocks can be rather large.

With so much sedimentary rock on Earth, one has to be a little bit curious as to where it came from and how it got here. If the Genesis Flood was a real catastrophic event, it would have torn up and removed huge amounts of sand, silt and small rocks and then transported them hundreds of miles from their native place. The sediments would then have been deposited into thick beds which hardened rather quickly. Some of the more famous sedimentary rock formations are: The Grand Canyon, Mt. Everest, The Colorado Plateau (over 130,000 square miles of sedimentary rock!), The Great Smokey Mountains, Ayers Rock in Australia (a huge sandstone formation over a 1,000 feet high), and Pamukkale (a large travertine formation in Turkey which is 8,100 feet long, 1,800 feet wide and 450 feet high). These types of sedimentary formations are not being formed like this today. Even the famous travertine terraces of Mammoth Hot Springs of Yellowstone National Park are not this big.

Travertine – a limestone sedimentary rock produced by the precipitation of chemical-rich waters that have permeated limestone rock from below ground

Chemical Sedimentary Rocks

Chemical sedimentary rocks have been deposited by supersaturated mineral-rich solutions. No one really knows how they were formed. One of the chemical sedimentary rocks is limestone. And as the name implies, the rock is made up of lime cemented together with either calcium carbonate (calcite) or silica (quartz). Limestone has the mineral calcium carbonate in its makeup. One way that geologists test for calcium carbonate is by dropping a tiny bit of strong acid on it. If it fizzes, then the rock contains calcium carbonate and more particularly there is a good chance it is limestone.

A question that is sometimes asked is whether the Genesis Flood could account for all the limestone formations in one single event? The answer is a resounding "yes". It is exactly what one would expect given the global nature of the Flood and its power to tear up huge amounts of rock and soil, living things, transport them and then deposit them into successive layers.

Biochemical Sedimentary Rocks

As the name implies, these are rocks made from the remains of once-living plants and animals. A good example is coal, made from the compressed remains of ferns, leaves and wood. Geologists do not know for sure how the vast coal beds around the world were formed. But coal has been formed in the laboratory using heat, plant material and pressure – and a relatively short period of time!

Bituminous Coal   **Lignite Coal** (soft coal)   **Anthracite Coal**
(metamorphosed hard coal)

Examples of continuous coal beds throughout the United States

Activity DD:

1. *How could massive coal beds form in today's environment?*
2. *In your notebook write out an explanation for the formation of coal beds using your Biblical geology framework. Answers in the Answer Key.*

Early coal mining, Williams River, West Virginia, 1930's.   Portrait by Finley Taylor

Take Quiz #10, pg. 103.

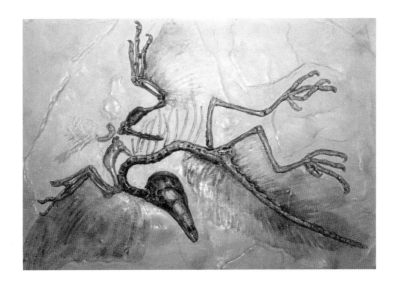

This is the classic "death pose": death by water catastrophe

III The Genesis Flood – The Historical Turning Point of Earth History

Section 5: Evidence of Creation and Catastrophe – The Fossils

Fossils and the Genesis Flood

Vocabulary: *Write out a brief description for each word in your notebook.*

Fossil (Fossilization)   Petrification   Mineralization   Fossil Cast   Fossil Mold
Invertebrate   Vertebrate   Trace Fossil

What causes something to fossilize? What conditions must be present to enable Soft body parts to be preserved? Scientists have tried to duplicate the process of fossilization with little success. However, observation would tell us that certain things must be present or absent if something is going to be preserved in stone:
1. The dead plant or animal must be free from oxygen. The presence of oxygen allows bacteria to grow and eat away at the dead plant or animal.
2. Therefore the plant or animal must be buried quickly.

3. Because scavengers quickly consume a dead animal, in order for complete skulls and skeletons to be preserved, rapid burial must also take place. Otherwise bones would simply dislocate and disappear.

4. The presence of the right chemicals that will aid in the fossilization process must be present. We know from mixing concrete that there must be a hardening agent of some kind or the mixture of mud, water and gravel will not set up and turn hard.

Maybe you have had a pet fish. You may have experienced the sad experience of having your pet fish die in the fish tank. What happened to the fish? Did it drift to the bottom of the tank and naturally get buried? Normally a dead fish will rise to the top of the tank where other fish have a meal of it. As the skin and tissues are eaten away or deteriorate, nothing remains to hold the bones together and the fish falls apart. No fossilization occurs. This is what we normally observe in the present. Now, if uniformitarianism is correct, and fossilization is a rare event in the present, then should there be many examples of fossils from the past? The fossil record, however, is rich, very rich with billions of exquisitely preserved fossils of plants and animals that once lived.

Activity EE: Study the pictures below and then ask yourself, "What event would have been able to account for this rich record of fossil plants and animals?"

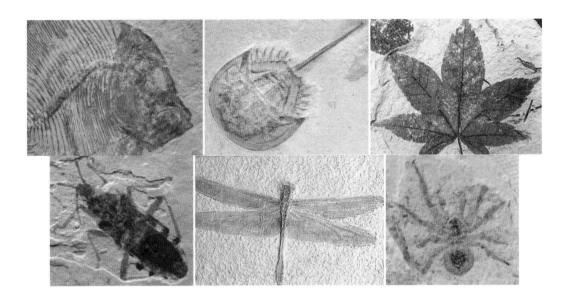

These fossils are **_not_** the exception! The past as recorded in the fossils we find today is so well documented that we can make out veins in wings, individual teeth, eye sockets, membranes, delicate spider parts and antennae. Such preservation was not natural, but sudden and catastrophic.

But, could small local upheavals and catastrophes have quickly buried plants and animals through millions of years of change and erosion? Yes, that could have happened. That is the idea behind uniformitarianism. Of course that is not testable, just as a global Flood is not testable by laboratory means. But what we notice, and this is the key to understanding the fossil record, is the preservation of billions of plants and animals preserved all over the globe and in large formations. For example, look at the map below and notice the vast amounts of coal formations around the world. These are not small isolated catastrophic burials, but huge amounts of coal formations over a wide geographic area.

Amazing detail was preserved in this fossil of a moth

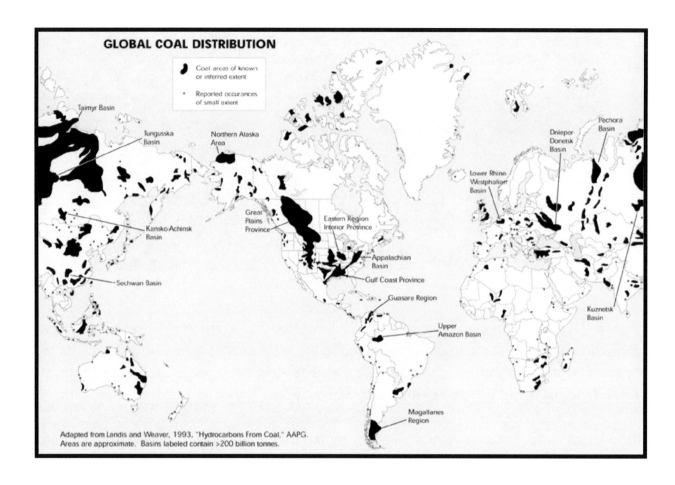

GLOBAL COAL DISTRIBUTION

Adapted from Landis and Weaver, 1993, "Hydrocarbons From Coal," AAPG.
Areas are approximate. Basins labeled contain >200 billion tonnes.

The amount of fossils that have been collected from around the world is enormous. Below is a list of the numbers of fossils housed in museums all over the world. This number does not include those fossils in private collections. If private collections would be included in this number, it would stagger the imagination. It would be huge! It boggles the imagination to think of just how this number of very will preserved fossil record could have been preserved by small isolated local geological events. It certainly is not happening that way today.

- Fossil invertebrates = 100,000,000 (animals without backbones; marine creatures)
- Fossil Insects: 1,000,000+
- Fossil Vertebrates: 2,718,000+ (animals with backbones)
- Fossil Whales: 2,000,000
- Fossil Fish: 500,000
- Fossil Bird Specimens: 200,000+
- Fossil Dinosaurs: 100,000
- Fossil Bats: 1,000+
- Fossil Turtles: 100,000
- Fossil Flying Reptiles (pterosaurs): 1,000
- Fossil Plants: 1,000,000+

The characteristic of the fossil record is replete with examples of once-living things that were caught in some kind of global watery catastrophe. From the fossils we have collected it becomes apparent that the fossils are a record of destruction and extinction, not evolution. No in-between-kinds (called transitional forms) have ever been found. If the evolution of life did occur, the lack of transitional forms would be highly improbable given the abundance of fossils found. Quite the contrary millions of fossils have been found and yet not one undisputed transitional fossil has ever been found.

What is a "transitional fossil"? Evolutionists today believe that all living things are related to one another. In other words life as we know it today developed from other creatures as they adapted and changed to become new and different creatures. In order to propose this as a legitimate explanation for the complexity and nature of life, we would need to see examples, lots of examples of plant and animal creatures in the process of becoming something else. Since there are no fossils that have been discovered to date that would support this idea, then we have to imagine just how living creatures would change from one type to a completely different type. Let's consider the following.

Activity FF: With each category of change listed below, write out what might have to be developed at each stage of life in order for there to be a complete change from one living thing to an entirely different living thing. Be sure to record the steps in your notebook. For our purpose, we will use the standard evolutionary chain. *Answers in the Answer Key.*

1. Evolutionists believe that life arose in the oceans around 550 million years ago. What would have to take place for the right environment, chemicals and structures to develop into the first multi-celled creatures?
2. Evolutionists believe that invertebrate animals gave rise to the fishes, vertebrate animals. What changes would need to take place in invertebrates to become vertebrate fish?
3. Evolutionists believe that fish gave rise to land creatures. What changes would need to take place in a fish for it to develop so that it could live on land?
4. Evolutionists believe that simple land creatures gave rise to the dinosaurs. What changes would need to take place for a simple land creature to become a dinosaur?
5. Evolutionists believe that dinosaurs gave rise to birds and mammals. What changes would need to take place in order for flight to develop? In order for egg-laying creatures to give rise to an animal that gives birth to live creatures?
6. As a matter of fact flight would have developed in four separate stages:
   - In insects
   - In birds
   - In reptiles – pterosaurs
   - In mammals – bats

7.   *What would need to change in order for flight to evolve in four separate living things at four separate times in evolutionary history and in four separate branches of life?*

For further study on this topic, the two books entitled, "Evolution, the Grand Experiment" and its sequel, "Living Fossils" would serve as an excellent unit study. See Appendix F for information and more ideas.

Notice the incredible details preserved in this fossil starfish. It must have taken extraordinary conditions for this to have taken place.

Take Quiz #11, pg. 103.

Glaciers in Antarctica – leftovers from The Ice Age

IV The Ice Age
Section 1: Evidence for an Ice Age

Vocabulary: *Look up and write out the definitions of the following words in your notebook.*

Volcanic Ash	Evaporation	Condensation	Glacier	Erratic	Arete
Glacial Trough	Glacial Cirque	Moraine	Horn	Relict	Loess
Glacial Til	Fluvium (Fluvial)	Striation	Pleistocene Epoch		Tarn
Plucking	Abrasion				

The subject of the Ice Age is usually quite unfamiliar to those who believe the Scriptures. First of all it is not mentioned in the Bible and so it would not normally be considered to be an important topic for discussion. So, why make it part of Biblical Geology here? Secular geologists teach that not only has there been an ice age, but there have been at least 30 of them spread out over millions of years. The last ice age supposedly began about 2 million years ago and ended around 12,000 years ago. This is known as the Pleistocene Epoch. If this is true, then the Scriptural account of a Creation and Global Flood cannot be true and the Bible is then shown to be a lie, because the Biblical chronology states very clearly that the Earth is around 6,000 years old. This is why we will consider the Ice Age at this point!

Is there evidence for an Ice Age? There is evidence that at some time in our past the northern hemisphere of the world and the higher latitudes were covered in very deep snow and ice. How do geologists know this?

Erratics – are boulders that are unnatural to the area in which they are found. One of the clearest examples is found in the northeastern part of Yellowstone National Park. The erratics are made of granite but the predominate rock of Yellowstone is volcanic rock, a completely different rock type.

The closest source of granite rock is the Beartooth Mountains located 50 miles northeast of Yellowstone. How did they get into Yellowstone National Park?

Glacial Troughs – Enormous "U"-shaped valleys that have been smoothed but not by water. Valleys or canyons that have been cut by water are "V"-shaped. The first picture below is of a "V"-shaped cut. The following two pictures are of "U"-shaped valleys.

These troughs, located high in the Beartooth Mountains, receive several feet of snow each winter, but it doesn't stay around all year. What event would have produced this much ice and snow to have carved these huge troughs? What would cause this much ice and snow to accumulate to such massive depths?   The two pictures above show a glacial trough over a thousand feet deep! Contrast these with the pictures below.

A "V" shaped canyon (a water gap) north of Gardiner, Montana.
The Grand Canyon is a "V" shaped canyon (a water gap).

Glacial Cirques – amphitheater shaped glacial valleys.

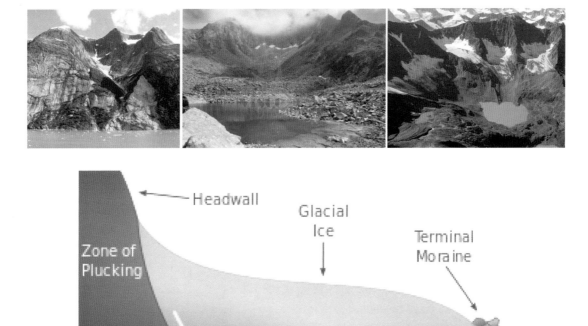

The above illustration begins to give you an idea of how these cirques come about.

Moraines – loose bits and pieces of rocks that have been broken off by moving glaciers and carried along as the glacier moves. The picture below is of Lake Wallowa in northeastern Oregon. It is a Ice Age glacial remnant. There are three types of moraines that you can see in the picture below:

1. Lateral moraines – piles of rocks heaped on the sides of glaciers at their end stages
2. Terminal moraines – piles of rocks left at the end of a glacier's route
3. Ground moraines – hummocky or bumpy small hills containing rocks moved by glaciers as they move across the ground

Horns - sharp peaks formed where the ridges separating three or more cirques intersect. The less resistant rock has been removed by moving glaciers.

The Matterhorn (Switzerland) – Pilot Peak and Index Peak (Montana)

Arete – a thin, almost knife-like, ridge of rock which has typically been formed when two glaciers eroded parallel U-shaped valleys.

Loess (pronounced, luss) – very fine wind-blown rock flour that accumulated during the end stages of the Ice Age. The picture makes these loess deposits look like rock, but it is not. The loess can be easily scraped away showing the fine flour-like texture. You have a sample of loess in your kit.

Pictures Courtesy USGS

Loess is found all over the world. The picture at the upper left of the page is of loess near Kansas City, Missouri. The picture in the upper right is from England. The maps, just above, show loess deposits all across the Midwest and across Alaska (orange shaded areas). Loess would be comparable to the dust piled high across the mid section of our own country during the dust bowl days of the 1930s.

Wind-blown dust of the "Dust Bowl" era during the 1930s

My mother-in-law's home in the midst of the Dust Bowl, Johnson, Kansas –
photo courtesy of Maxine Burkhard

Activity GG: *Study the photos below. How many glacial features can you identify? Number and record your answers in your notebook.* *Answers in the Answer Key.*

Take Quiz #12, pg. 103.

The Woolly Mammoth: Icon of the Ice Age
Millions disappeared sometime after the Flood, toward the end of the Ice Age

IV The Ice Age
Section 2: Where in the Bible is the Ice Age?

Now that we have learned about some of the features of the glacial activity of the past, let's explore the answers to the following questions:
   1) How does an Ice Age fit into the Biblical view of Earth history?
   2) How would this Ice Age have originated in a Biblical geology framework?

The Biblical chronology leaves no doubt that the Earth is about 6,000 years old. A global flood massively scarred the Earth about 4,500 years ago and man spread out from Babel about 100 years after the Flood. Where do we fit an Ice Age? There isn't a whole lot of room for a 2 million year old ice age! The Genesis Flood is the key to this mystery.

Read Genesis 7:11. Part of the breaking up of the fountains of the great deep would explain a source for much of the water that flooded the Earth. But it also explains volcanoes and the release of magma from deep under the ocean. The magma released into the ocean waters being forced out of their deep aquifers would have warmed the waters and increased the evaporation. The volcanic ash and aerosols released into the atmosphere from thousands of erupting volcanoes would have blocked much of the reflective sunlight making the atmosphere much cooler. Increased moisture from warmer oceans coming into contact with cooler air would condense in the form of a lot of snow – lots and lots of snow. As the snow would compress with the increase weight of more snow, ice sheets would form. The temperature would actually be quite mild. You may have experienced this as you have seen big flakes of snow falling in the winter months. Temperatures are actually quite mild while snow is falling. Only after the snow stops does it start to get colder.

As the Earth began to settle down after the Flood and its aftermath, volcanic eruptions would subside. The sun would shine through and the snow and ice would begin to melt producing Ice Age flooding. At this point summers would become warmer and winters colder. Dryer air would accompany the climatic changes producing strong winds and blowing loess. This is likely how many of the woolly mammoths died. Many of these creatures have been found buried in glacial loess.

In summary the Genesis Flood with its catastrophic tectonic upheaval and volcanic eruptions would have caused a rapid build up in snow and ice and then a rapid melting and catastrophic flooding. This is the explanation for the Great Missoula Flood and the Scablands National Monument in Washington State.

Could volcanic eruptions really be the catalysts to such an event as the Ice Age? Let's take a look at two examples in recorded history – Kraktau and Mt. Tambora.

Krakatau – 1883 and Mt. Tambora – 1815

As the volcano Krakatau erupted in 1883 it put hundreds of tons of volcanic ash and aerosols into the atmosphere. One encyclopedia states, "These acidic aerosols (ash clouds) sufficiently blocked enough sunlight to **drop the Earth's temperature by several degrees for a few years**. Their presence in the atmosphere also created spectacular effects over 70% of the earth's surface."

Mt. Tambora in Indonesia erupted in 1815. It has been called the largest volcanic eruption in recorded history. One newspaper wrote, "Tambora erupted in 1815 killing 92,000 people. 1816 became the year without a summer as the global climate effects were felt. Aerosols from the Tambora eruption blocked out sunlight and **reduced global temperatures by 3 degrees centigrade...** Europe missed a summer, and India had crop failures...On June 4, 1816 frosts were reported in Connecticut, and by the following day, most of New England was gripped by the cold front. On June 6, 1816 snow fell in Albany New York, and Denneysville, Maine...Such conditions occurred for at least three months and ruined most agricultural crops in North America. Canada experienced extreme cold during that summer. Snow (12 in) deep accumulated near Quebec City from June 6th-10th, 1816..."

Just how big were these eruptions? The chart below begins to give you an idea by way of a comparison with the eruption of Mt. St. Helens.

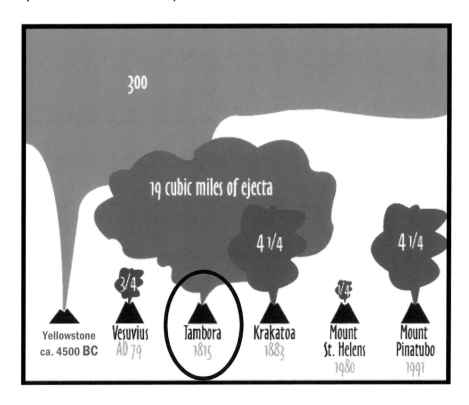

The one mechanism not present during these volcanic eruptions, however, was the warmer ocean caused by the magma that must have been produced by the breaking up of the fountains of the great deep recorded in Genesis 7:11. The Genesis Flood was a unique historical event and according to Genesis chapter 9, will never again be repeated. The Genesis Flood provided these 2 mechanisms that would be powerful enough toward the end of the Flood to produce an Ice Age. Geologists have advanced over a 100 explanations for what could have possibly brought on an Ice Age, and none of them have been totally accepted. Something has been missing in each explanation – the **combination** of the mechanisms of the warmer oceans (increased evaporation) and the cooler atmosphere brought on by the multiple volcanic eruptions.

Artist's rendition of the eruption of Mt. Tambora and a map location

Take Quiz #13, pg. 103.

Take Comprehensive Exam, pg. 109.

Appendix A – Radiometric Dating and the Age of the Earth

Radiometric Dating and the Age of the Earth

The Bible's chronology makes it clear that the Earth originated about 6,000 years ago by the word of God and that a global catastrophic flood destroyed the Earth about 4,500 years ago. Secular geology claims that the universe is 15 billion years old and the Earth is about 4.6 billion years old. If this is true, then the entire Bible is a lie! Furthermore secular geologists claim to have proof positive that the Earth is 4.6 billion years old by way of radiometric dating. Let's look at this and see if it is indeed proof positive of an old Earth.

Radiometric dating

What is radiometric dating? Geologists claim that it is a type of natural clock. It is based on the fact of radioactivity; that some elements are unstable and therefore "decay". They are radioactive. Furthermore it is claimed that this decay occurs at a known, constant rate. It is a fact that decay can be measured in the present. So, according to geologists, all one has to do is to keep track of this present decay and then calculate how long it would take for a radioactive element to decay completely. An unstable element is one that has too many neutrons in its nucleus. Look at the illustration of the Carbon atom below.

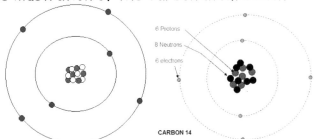

The normal Carbon12 atom (6 protons, 6 neutrons) on the left, and
The abnormal (radioactive) Carbon14 atom (6 protons and **8 neutrons**)
The number of electrons equals the number of protons

A radioactive element is called an isotope (meaning "same place") because it occupies the same place as the normal atom on the Periodic Table, but has a different nucleus (too many neutrons; too much energy; too many protons).

The Periodic Table

The Periodic Table was developed by a Russian named Dmitri Mendeleev in 1869. He developed his table to illustrate periodic trends in the properties of the then-known elements. Mendeleev also predicted some properties of then-unknown elements that would be expected to fill gaps in this table. Most of his predictions were proved correct when the elements in question were subsequently discovered. The Table is organized on the basis of atomic numbers, electron configurations, and recurring chemical properties. Elements are presented in order of increasing atomic number (number of protons). There are 18 columns. The columns are called groups. The groups are based on atomic structure of the elements; all elements in a group have similar atomic structure.

Periodic Table of the Elements

1	2	3	4	5	6	7	8	9	10	11	12	13	14	15	16	17	18
1 H 1.01																	2 He 4.00
3 Li 6.94	4 Be 9.01											5 B 10.81	6 C 12.01	7 N 14.01	8 O 16.00	9 F 19.00	10 Ne 20.18
11 Na 22.99	12 Mg 24.30											13 Al 26.98	14 Si 28.09	15 P 30.97	16 S 32.07	17 Cl 35.45	18 Ar 39.95
19 K 30.10	20 Ca 40.08	21 Sc 44.96	22 Ti 47.88	23 V 50.94	24 Cr 52.00	25 Mn 54.94	26 Fe 55.85	27 Co 58.93	28 Ni 58.69	29 Cu 63.55	30 Zn 65.39	31 Ga 69.72	32 Ge 72.61	33 As 74.92	34 Se 78.96	35 Br 79.90	36 Kr 83.80
37 Rb 85.47	38 Sr 87.62	39 Y 88.91	40 Zr 91.22	41 Nb 92.91	42 Mo 95.94	43 Tc (97.91)	44 Ru 101.07	45 Rh 102.91	46 Pd 106.42	47 Ag 107.87	48 Cd 112.41	49 In 114.82	50 Sn 118.71	51 Sb 121.75	52 Te 127.60	53 I 126.90	54 Xe 131.29
55 Cs 132.91	56 Ba 137.33	57 La 138.91	72 Hf 178.49	73 Ta 180.95	74 W 183.85	75 Re 186.21	76 Os 190.23	77 Ir 192.22	78 Pt 195.08	79 Au 196.97	80 Hg 200.59	81 Tl 204.38	82 Pb 207.2	83 Bi 208.98	84 Po (208.98)	85 At (209.99)	86 Rn (222.02)
87 Fr (223.02)	88 Ra (226.03)	89 Ac (227.03)	104 Rf (261.11)	105 Ha (262.11)	106 Sg (263.12)												

58 Ce 140.12	59 Pr 140.91	60 Nd 144.24	61 Pm (144.91)	62 Sm 150.36	63 Eu 151.97	64 Gd 157.25	65 Tb 158.93	66 Dy 162.50	67 Ho 164.93	68 Er 167.26	69 Tm 168.93	70 Yb 173.04	71 Lu 174.97
90 Th 232.04	91 Pa 231.04	92 U 238.03	93 Np (237.05)	94 Pu (244.06)	95 Am (243.06)	96 Cm (247.07)	97 Bk (247.07)	98 Cf (251.08)	99 Es (252.08)	100 Fm (257.10)	101 Md (258.10)	102 No (259.10)	103 Lr (262.11)

92 U 238.03 — The top number is the atomic number. It tells the number of protons (and electrons) in the element. The middle letter(s) is the chemical symbol for the element. The bottom number is the atomic mass. It tells the total amount of protons and neutrons in the element. In this case U238 has 146 neutrons.

Radioactivity

"Radioactivity is caused when an atom, for whatever reason, wants to give away some of its energy. It does this because it wants to shift from an unstable configuration to a more stable configuration. The energy that is released when the atom makes this shift is known as radioactivity. In other words, radioactivity is the act in which an atom releases radiation suddenly and spontaneously." *Tec-Fac Web site.* Some radioactivity is extremely dangerous. Some is not. The important thing to remember is that radioactivity can be measured. The unit of measure of "decays" per second is called the Becquerel, after Henri Becquerel, its discoverer.

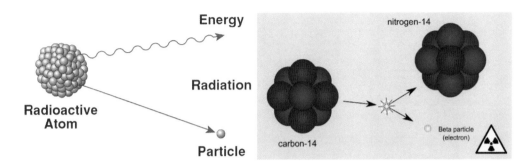

Radioactive decay

If radioactive elements (atoms) lose particles in the decay process, what happens to the original radioactive element? Obviously there is going to be a change that takes place. The radioactive element theoretically becomes something else. I say theoretically, because the decay process can require an immense amount of time to complete and would not be able to be observed throughout the entire process. The unstable element that is decaying is called the "parent" and the stable element into which it theoretically eventually changes is called the "daughter". Many of the radioactive elements used in radiometric dating require huge amounts of time and no one has recorded or witnessed the entire decay process. For example the time it takes for ½ of Carbon 14 to decay, measured at present measurable rates is a little over 5,000 years. Then in another 5,000 years ½ of what is left decays and so on until it is all gone. Has anyone ever observed the complete process to know whether it has actually completed this process or not? How could they? Below is an illustration of the idea of half life. Theoretically all that is left at the end of the decay process is the stable daughter element.

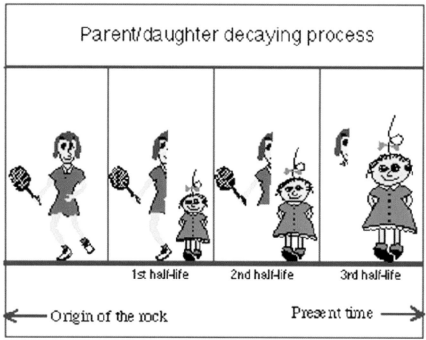

Parent/daughter decaying process

1st half-life   2nd half-life   3rd half-life

← Origin of the rock   Present time →

Courtesy of Minerals.cr.usgs.gov

The half life of the radioactive element is determined by the present Becquerel. The time is then projected forward or backward to arrive at a time when the process theoretically started or when it would theoretically end. In our Carbon14 atom above, the stable element, into which C14 would decay, is Nitrogen14. This is a chemical prediction based on what we know of present radioactive decay. Remember that the process, if totally accurate, would take thousands of years to complete. No one has been around that long.

Uranium decay

Radioactive uranium238 (U238) is a very heavy element. But notice its half life from the picture below! So the question is, has anyone ever observed this? The answer is obviously "no". Then how do we know that it operates this way? We can guess based on the chemistry of radioactive decay. It is a matter of statistics. If U238 were to continue to lose particles like it seems to do in the present, then over such and such a length of time it will go through a "chain" of decays until a stable element is arrived at – in this case, Lead206.

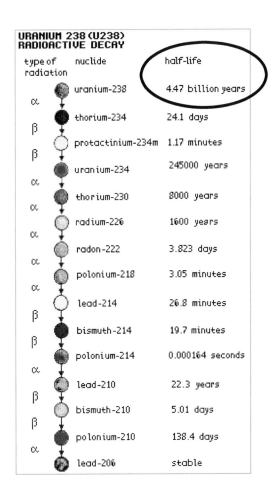

U238-Lead206 is a commonly used method in calculating the radioactive decay/age of a rock. But if no one was around to actually test or record the entire process from beginning to end, then how do we know this process works? We don't! Geologists make **assumptions** based on the present observation of radioactivity. Does that sound familiar? It should. It is an application of the philosophy known as uniformitarianism – *the present is the key to the past.* So, by assuming the present has always been true in the past, geologists can use the process of radioactive decay to date rocks. How do they do this?

Only rocks that are thought to have been molten at one time are considered to be valid specimens. So, fossils cannot be directly dated because of contamination. Fossils are dated by another means, also by way of uniformitarianism. That is, creatures change in the present. Therefore creatures have changed in the past and given enough time, have changed radically into other types of creatures – evolution. Look at the column below. It is presented as a scientific fact. But in reality it is a hypothetical idea. The column in its entirety has not been found anywhere on the Earth. Some of the rock layers do occur in an apparent order, but there is also an explanation for this apparent order that modern geologists reject. It is a general order (with many exceptions) produced by the sorting action of the Genesis Flood. If the fountains of the great deep burst open first, as Genesis states, then one would expect that marine (sea) creatures would have been the first victims of the Flood. As the Earth became inundated with water, other creatures would have followed as the Flood swallowed them up. Man, being the most mobile, would have most likely been buried last, if at all. Creatures swimming in order to survive would have ultimately drowned and their remains would have been eaten by fish or sharks. The apparent order in the rock layers has been assumed by geologist to be the result of an evolutionary process, having rejected the idea of a global flood.

The Geologic Column or Time Scale or Time Table
As the idea of change in creatures expanded to an evolutionary view where one type of creature was thought to have changed into an entirely different creature, so the amount of time for nature to accomplish this was also necessarily expanded from several thousand years to 550 million years by the end of the 19th Century. The basic Geologic Column as we have today was set by the mid to late 19th Century. In other words the concept of time plus evolution was a philosophical shift from a Divine Creation taught in the Book of Genesis to a totally naturalistic and atheistic one taught through uniformitarianism. Today many people think that the Geologic Column and the millions of years it involves is a scientific fact supported by radiometric dating. Remember that radioactivity was not even discovered until after the Geologic Column had been in place for many years. Most people are unaware of this. It was not science that formulated the Geologic Column, but philosophy. It was a naturalistic attempt to explain the apparent order of fossils. That is, marine fossils on the bottom of layers, dinosaurs closer to the top. It is an idea, and an accepted idea, but not proven science.

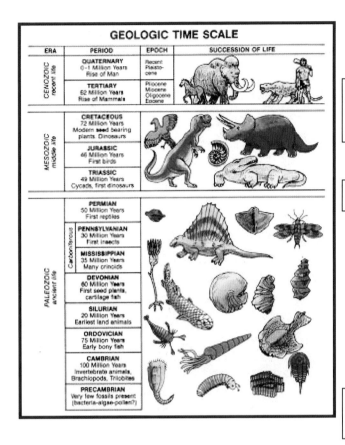

GEOLOGIC TIME SCALE			
ERA	PERIOD	EPOCH	SUCCESSION OF LIFE
CENOZOIC recent life	QUATERNARY 0-1 Million Years Rise of Man	Recent Plesitocene	
	TERTIARY 62 Million Years Rise of Mammals	Pliocene Miocene Oligocene Eocene	
MESOZOIC middle life	CRETACEOUS 72 Million Years Modern seed bearing plants. Dinosaurs		
	JURASSIC 46 Million Years First birds		
	TRIASSIC 49 Million Years Cycads, first dinosaurs		
PALEOZOIC ancient life	PERMIAN 50 Million Years First reptiles		
	PENNSYLVANIAN 30 Million Years First insects	Carboniferous	
	MISSISSIPPIAN 35 Million Years Many crinoids		
	DEVONIAN 60 Million Years First seed plants, cartilage fish		
	SILURIAN 20 Million Years Earliest land animals		
	ORDOVICIAN 75 Million Years Early bony fish		
	CAMBRIAN 100 Million Years Invertebrate animals, Brachiopods, Trilobites		
	PRECAMBRIAN Very few fossils present (bacteria-algae-pollen?)		

Extinction Event
65 Million Years
Ago

Extinction Event

550 Million Years
Old

How is a radiometric date arrived at?

Let's start with a volcanic rock. And let's say it is chemically analyzed to contain 80 parts of uranium and 20 parts of lead. Is the rock young or old? Since more of the uranium is present than lead, then the rock must be young. If the ratio was reversed, then the opposite would be true – *given the following assumptions*:

1. We *assume* that the initial state of the rock started with a certain amount of uranium and no lead. In other words the initial state of the rock is *assumed*.
2. We *assume* that there was no lead present at the start of the process.
3. We *assume* that no uranium came from some other source.
4. We *assume* that all of the lead that is present came from the decay of uranium and that it did not come from some other source.
5. We *assume* that the decay rate or process has not been interfered with from some other means or sources.

So long as we go with these *assumptions*, we can statistically figure an age for the rock.

But what happens when different ages are obtained, as in the case of Mt. St. Helens lava, which was known to be ten years old, but which dated in the hundreds of thousands of years. Again, geologists simply *assume*:

1. You or the laboratory contaminated the samples.
2. You or the laboratory made a mistake in calculations.

But what if the sample dates different ages using different dating methods, as in the case of rocks from the Grand Canyon where different dates for the same rock were obtained using different methods. Then the "umpire" is the Geological Time Scale (Geologic Time Table). Since dates back to 550 million years old were worked out in the 1800s, and have already been agreed upon, if the date for the rock appears to be too old or too young, then the dates are either thrown out or selected. The date which is the closest to what the geologist thinks it is becomes the accepted date. *And that's radiometric dating.*

Radiometric assumptions are not reliable

What might have caused radioactivity to change in the past? The Genesis Flood would have added a tremendous amount of heat to rocks. It would have also added a tremendous amount of hot water to the rocks. And in fact it has been shown that radioactive decay was rapidly sped up in the past in a number of substances:

1. Carbon14 was discovered in diamonds thought to be at least several billion years old. According to present measurements of the decay of C14, there should be absolutely no detectable C14 after around 100,000 years.
2. Carbon 14 was discovered in petrified wood thought to be at a few hundred million years old.
3. Recent observed lava flows in New Zealand and in Hawaii dated excessively old according to modern radiometric dating methods.
4. In addition if radiometric dates need to be checked by another source, the Geologic Time Table, for accuracy and reliability, then are they reliable methods?

These examples show that the radioactive decay process can be affected by Earth's processes, although geologists do not know why or how yet. Flood geologists suspect that it was by way of a unique historical global geological event we call the Genesis Flood. If this is so, then thousands if not hundreds of millions of years worth of radioactive decay occurred in a matter of a few months at some time in Earth's past. The radioactive process has been dynamically affected. This means that the radiometric clocks that geologists have developed are really not reliable. Even though radioactivity is an observable fact, the measurement and predictions of this process have been affected by something else and should not be used as a basis for age/dating. It would be like hanging on to a clock or watch that ticks, but is always gaining or losing time. Why depend on it? It's a clock, but is it reliable?

Why do geologists continue to use an unreliable method?
Geologists continue to use the radiometric dating methods in spite of the obvious flaws. Why? Although there is rarely an open public admission of these errors, the radiometric dating methods provide the only "scientific" alternative to the young Earth conclusions of the Bible. At the center of the radiometric dating rational are the influence of The Enlightenment and consequently the extremely narrow view of uniformitarianism. Anything but the Bible!

(From top left clockwise) Mt. St. Helens recent dacite dome, Mt. Ngaruhoe in New Zealand and recent lava flow in Hawaii

Appendix B – An Introduction to Biblical Geology, PreK-4th

Dear Parents, people who believe the Bible are ridiculed when it comes to things like the Creation and the Genesis Flood. I have seen many Christians compromise their trust in the Bible because of what people say. While we wrestle with our faith at times, think of what your child will go through when he/she is older. We must start now to teach them about how to view the earth from beginning to end. The Bible is more than a philosophy for life. Much of the Bible is recorded history supervised by the One who was there and saw it all. Teaching our kids to study the rocks in light of what God has spoken, will give them a solid foundation for trusting God's word in the future as they are challenged again and again. As you progress through the different learning levels of your kids, you can build off of the material presented here. Kids are naturally drawn to rocks. Let's build off of that interest by telling them the whole story – not just the Sunday morning portion of it.

There are a few things that a child of this age should be made aware of and that will serve as the foundation for the development of a Biblical view of Earth history. That is what is covered in the Grades Prek-4th.

The word geology comes from two Greek words – *geo*, meaning earth, and *logi*, meaning the study of. So, geology is the study of the earth. When we study the earth, we should be curious about who made the earth, how and why it was made, all the rocks we see and, the future of the earth. Aren't you a little curious about these things?

But, in order to answer these questions, we need to know someone who was actually there when the earth was made. Let's see. Who could that be?

Think back to your favorite birthday party? Were you there? Who else was there? What kinds of things did you do? *Make a list of the things you remember:*

_____

_____

The things you listed are called "evidences". These are additional things that help me have trust in what you say.

Now, I wasn't there when you had that fun birthday party. How will I know about your birthday party? Well, you told me so. I believe you because you told me. And, I can also believe you because of the "evidences" you listed above.

The same would be true of your parent's marriage. You know all about their marriage because they told you and because of the pictures and the reports from others who saw their marriage. Were you there? No, but others were. And so, you believe them.

The same is true when the earth was made. You weren't there. Your parents weren't there. Your friends weren't there. Your teachers weren't there. But God was there. And He has told us so in His Bible.

Activity: Memorization: "In the beginning God created the heaven and the earth." *Genesis 1:1*

So, God was there when the earth was made and He has told me so in His Bible. Here is a great little saying that I learned early on in my Christian life: *"God said it. I believe it. That settles it."* This is what faith is all about. Faith is trusting the word of the One who was actually there when the earth was made.

But, just like you had evidence of your birthday party, so God has left us with evidence that He made the earth. *Can you think of evidence that God has left us that He made the earth?*

_____

_____

Now that we have begun a good habit of listening to what God said about how the earth was made, do you think God has anything to say about the rocks that we see around us?

Of course God's main interest is in people, isn't it? How do we know this? Well, God's Bible tells me so.

Activity: Memorization: *John 3:16*

OK. So, where did all the rocks come from? Remember, rocks are part of the evidence that God left us about what He did when He made the earth. Some rocks are pretty and they show the work of God Who is a Master Artist. Some are useful to us and therefore show God's care for us. Some rocks are just interesting, like God is. I think rocks from volcanoes are especially interesting. *What are some of your favorite rocks?*

_____

_____

So, some rocks show something about our God as our Maker. But some rocks, although interesting to look at, seem to tell a different story. Some rocks are twisted and crushed. Can you find a rock that looks like this?

Can you think of a story from God's Bible that might tell us how these rocks got this way?

_____

_____

If you guessed the story of Noah's Flood, you would be correct. Let's get God's Bible and read about Noah and his life from *Genesis chapters 6-8.* Now, remember, God was there when the earth was flooded. So, we can trust what He says about this event. *When you have finished listening to the story, list all the phrases that might tell us something about what happened to the earth during Noah's life.*

_____

_____

_____

So, some of the rocks we see around us will definitely tell us a story about God, our Maker. But, most of the rocks around us will tell the story of how God destroyed much of the earth He had made. *Why did He do this?*

Activity: Memorization: *Genesis 6:5*

To me, this is one of the saddest verses of God's Bible. God made a beautiful earth, but because of the evil of man, God had to destroy it. And much of the earth we see today is evidence of God's judgment. The rocks today tell us that God is watching man and will hold him responsible for what he does.

Even most of the dinosaurs died as a result of man's evil. How do we know this? Because the fossils are evidence of this. Fossils are the remains of once living plants and animals preserved in the rocks.

Have you ever found a fossil? Fossils tell us how powerful the Flood of Genesis was. This was no ordinary flood! This Flood had never happened before and it has not happened since. Most fossils I have found tell me a story that matches what I read in God's Bible.

Activity: Memorization: *Genesis 7:23*

Another piece of evidence about the Flood that we can see today is found in all the volcanoes that are around the earth. Some volcanoes have stopped erupting and only their mountains are left. But some are still erupting. Where did all the volcanoes come from? I think I know. Let's see if you can guess? Read this verse from God's Bible.

Activity: Memorization: *Genesis 7:11*

So, it just didn't rain for forty days and nights, but the whole earth was torn apart. Huge earthquakes must have taken place. Many mountains must have crumbled. Tremendous amounts of magma (lava) must have shot up from below the earth to form volcanoes. *Can you think of other things that must have happened with so much earth movement?*

All around the earth we can still see what has been left of the Flood. The earth is weak in many places. Earthquakes, tsunamis, tornados, volcanic eruptions are all a part of the weakened earth as a result of the Flood. God helps us when we are affected by these things, but He has left them as evidences of the tremendous Flood that once destroyed the Earth.

This is Biblical Geology! The rocks all around us tell a story of destruction and awesome power, like what God's Bible tells me. Some people may want to tell a different story, but they weren't there, were they? Who would you rather trust? God Who was there, or the people who were not there, but tell a different story?

Appendix C
Quizzes

Part I – The History of Modern Geology
Section 1 and 2 – Quiz #1
1. Who popularized the idea of uniformitarianism?
2. Who was the Father of Modern Geology?
3. As early as the 1600s this geologist sought to interpret the rock layers in light of the global flood of Genesis.
4. The movement during the 1700s that was characterized by rejection of the Scriptures as an explanation for the past history of the Earth.
5. This so-called "scientific" view seeks to explain everything in our world without the aid or interference of a God. It is the same thing as atheism.
6. "The present is the key to the past"
7. What is Deism?
8. The construct developed by secular geologists in the 1800s that interprets all geological data.

Part II – The Origin and Nature of the Earth
Section 1 – Quiz #2
1. What was the one thing that bothered Albert Einstein?
2. Briefly list the order of creation from *Genesis* Chapter One.
3. What is a "framework"? How does Genesis serve as a framework for understanding geological data?
4. Why is the possibility of life arising by chance zero?

Section 2 – Quiz #3
1. What is the study of the universe? Why did the Greeks choose this word?
2. Name the two most abundant elements on Earth.
3. What are the three parts of the atom? What is the nucleus?
4. What makes man different from all other created things?
5. Name two lighter colored minerals. Name two darker colored minerals.
6. What are the 12 rock-forming minerals?
7. What are the 9 mineral families?

Section 3 – Quiz #4

1. What two elements make up water?
2. According to 2 Peter 3:5, from what substance was the Earth formed? How does this differ from the Big Bang idea?
3. How much of the Earth is covered by water? What part does water play in the maintenance of our Earth?
4. Briefly describe the pH scale. What does this tell us about the design of the Earth?

Section 4 – Quiz #5

1. What are the two main elements that compose the Earth's atmosphere?
2. What two purposes does the Earth's atmosphere serve?
3. How does the Earth's magnetic field help life on Earth?

Section 5 – Quiz #6

1. Geologists think the Earth's crust has two parts. What are they and what main rocks seem to characterize each part?
2. What are plutonic rocks?
3. Name two plutonic rocks.
4. Since no one has seen plutonic rocks form, how would you explain their origin from a Genesis framework?

Part III – The Genesis Flood
Section 1 – Quiz #7

1. Briefly list the parts of the Genesis Flood as told by the Scriptures.
2. How would you describe the Bible?

Section 2 – Quiz #8

1. What are the two stages of the Genesis Flood?
2. Give one geologic evidence for each stage of the Flood.

Section 3 – Quiz #9

1. How would the Genesis Flood explain each of the following geologic landforms:
    1. Folded sedimentary rocks
    2. Devil's Tower
    3. Planation surfaces

4. Seamounts

Section 4 – Quiz #10
1. Name two volcanic rocks.
2. Name two metamorphic rocks.
3. Name two sedimentary rocks.
4. Briefly describe the differences among the rock types above.
5. What is a "clast"?
6. What are chemical sedimentary rocks?
7. Discuss how limestone would be evidence for a global flood?

Section 5 – Quiz #11
1. Name two conditions necessary for fossilization.
2. How do fossils demonstrate the truth of Genesis chapter One?
3. What is a transitional fossil? Why are these necessary for the evolutionist?
4. What is the evolutionary order of life and does it contrast with the idea of creation as taught in Genesis chapter One?

IV – The Ice Age
Section 1 – Quiz #12
1. Name three geologic features that show that there was some kind of "ice age" in the past.
2. What is loess? How could it account for the preservation of mammoths?

Section 2 – Quiz #13
1. Briefly describe where the idea of an ice age would fit into the Bible's chronology.
2. What are the two main stages of an ice age?
3. How would the Genesis Flood framework explain the onset of an ice age?
4. What are the two main ingredients necessary to begin an ice age?

Appendix D – Activity Answer Key

Activity A

1. Nicholas Steno was *Nicolas Steno* (1638-1686) was a geologist who believed in the Biblical revelation of a 6 day creation and that the rock layers and fossils were the result of the global flood of Genesis chapters 7-9. He is considered to be one of the founders of the science of *stratigraphy.*

2. Naturalism is an atheistic belief. It is not science. It is a belief that the universe and all that is in it is here without the aid or interference of a god.

3. *"The present is the key to the past".* In other words geologists think they can gain an understanding about Earth history by only studying present geological forces and processes.

4. The answer should include at least some of the following: Modern geology insists that the age of the Earth can only be answered by science; modern geology insists on making the argument one of science versus religion; modern geology refuses to accept any conclusions made by creation scientists as valid. Consequently the public has been led to believe that modern geology has the final authority and say on the age of the Earth.

5. Modern geology and Biblical geology are diametrically opposite *beliefs* and equally as *spiritual.* Both are attempts to tell a story of Earth history.

Activity B

Answers could include: Uniformitarianism assumes Biblical revelation is not legitimate. It assumes that events like the Biblical Flood could not have happened. It assumes that since God is not science, therefore Genesis 1 could not have happened.

Activity C

1. Answer vary with observations.
2. Answers could include: It is scientific in that it analyzes chemical properties of rocks and minerals. It is religious in that it assumes an atheistic origin of the rocks and minerals.
3. Answers could include: Ostracism, criticism, mockery, loss of job.
4. It mixes belief about the past with some science.

Activity E

1. Day 1: God created the heavens and the earth, light, separated the light from the darkness, named the day and the night: Day 2: Made the expanse, placed waters above and below it, named the expanse heaven; Day 3:

Gathered the waters, dry land appeared, named dry land earth, gathered waters He called seas, vegetation; Day 4: Lights to separate the day from the night, and to create signs, seasons, for days and years; and to give light on the earth, two great lights, one to govern the day, the other to govern the night; Day 5: Sea creatures, birds created; Day 6: Land creatures created, man created (male and female, he created them.)

2. 1) Day 4; 2) Day 5; 3) Day 5; 4) Day 6; 5) Day 6; 6) Answers could include: The chart in this section gives some of the many differences.

Activity G

1. a) answers will vary, but could include: God gave Adam and Eve a command and they disobeyed it which brought about corruption, decay and death into the once-perfect creation of Genesis chapters 1-2.
   b) Answers could include the following: through one man sin entered the world; corruption entered the world, i.e., thorns and thistles, enmity, death

2. a) Established the earth, covered it with the deep, waters fled, mountains rose, valleys sank, set a boundary for the waters
   b) Psalm 104:5

3.
   a) Hebrews 1:2,10; He created it, He made it
   b) Hebrews1:2,10; the Son, the Lord
   c) Hebrews 1:3,
   d) Deism would not continue to have any god involved, once the creation aspect was done.

4. We read the Bible because it is the only record we have of how God created the world. In addition, it is revelation, which means that God revealed the past to us.

5.
   a) God
   b) That God created the world in six days, and rested on the seventh

Activity J

Answers could include: Salt is good because Jesus said it is good. It is good for preservation, for the proper functioning of the body, and for taste. It is potentially bad when it is consumed in unhealthy amounts. When it is refined, it loses some of its properties.

Activity L

Possible answers could include: purification and nutrition for both plants and animals.

Activity R

1. Answers vary.
2. Answers could include: Canyons are similar to canyons on land. Mountains are similar to mountains on land. Volcanoes are similar to volcanoes on land. Basalt lava is abundant, as it is in places like the Columbia Plateau of Washington state, or Hawaii.

Activity U

1. Death is not mentioned.
2. None of these are mentioned.
3. Answers might include: death, destruction, corruption, decay.

Activity V

Answers might include: An explanation of why things wear out or decay; They could explain why, with time, things are not developing, but decaying.

Activity W

1. Day 1: God created the heavens and the earth, light, separated the light from the darkness, named the day and the night;
2. Day 2: Made the expanse, placed waters above and below it, named the expanse heaven;
3. Day 3: gathered the waters, dry land appeared, named dry land earth, gathered waters He called seas, vegetation;
4. Day 4: lights to separate the day from the night, and to create signs, seasons, for days and years; and to give light on the earth two great lights, one to govern the day, the other to govern the night;
5. Day 5: Sea creatures, birds created;
6. Day 6: Land creatures created, man created (male and female, he created them.)
7. It says, "...evening and morning, the first day," and so on.
8. God told Moses to rest on the seventh day, because he had created the world in six days, and rested on the seventh.

9. The earth; The Big Bang says galaxies came first, then earth; God placed the sun, moon and stars into space.
10. Vegetation, plants yielding seed, fruit trees bearing fruit after their kind, the great sea monsters, every living creature that moves, every winged bird after its kind, cattle, creeping things, beasts of the earth after their kind,man.
11. They could be beasts of the earth or creeping things.

Activity Y

1. Answers could include: "...blot out from the face of the earth every living thing that I have made." "The water prevailed so that all the high mountains everywhere under the heavens were covered." "All the flesh that moved on the earth perished: birds, cattle, beasts, every swarming thing, and all mankind." "All that was on the dry land died." 'He blotted out every living thing that was on the face of the land, from man to animals to creeping things and to birds, they were blotted out from the earth." "And the water prevailed upon the earth 150 days."
2. Answers could include: All land animals and birds died. All people died (except Noah and his family), the surface of the earth was totally rearranged.
3. Answers could include: fountains of the great deep burst opened, floodgates of the sky were opened, rain fell up on the earth 40 days and 40 nights. All the high mountains everywhere under the heavens were covered. Water receded steadily from the earth.
4. Adam to Noah – about 1600 years. The Flood took place around 4500 years ago.

Activity Z
The bread should bend.   The sticks should break or splinter.

Activity AA
Something very powerful had to have sheared these structures flat. Normal erosion produces valleys, cracks, and other irregularities.

Activity CC
After two days, the mixture should be hard. What this activity demonstrates is that it takes the right mix of the right chemicals and sediments to make sedimentary rock, not a lot of time.

Activity DD

1. They couldn't, because wood and peat would rot before they could fossilize.
2. Answers could include: Catastrophic upheaval of plants, rapid burial of plants, pressure and heat generated by the tectonic activity of the Flood.

Activity FF

1. The environment would have to be perfectly suited for each developing individual creature or plant at every stage of development. There is absolutely no way that these two things (the environment and evolving creatures) could have been perfectly coordinated so that what develops just happens to be perfectly suited.
2. Answers could include: development of a backbone, gills, fins, sight, smell all perfectly coordinated to work together.
3. Answers could include: development of lungs, special sight, legs, special eggs, special digestion all perfectly coordinated to work together.
4. Answers could include: development of complex lung structure, cold or warm blooded system, hip structure, digestion all perfectly coordinated to work together.
5. Answers could include: (a) development of feathers, the right bone structure, the right lung structure; (b) development of hair, development of a different reproductive system, development of the right birthing system, digestion, mammary glands all perfectly coordinated to work together.
6. Answers could include: The answer to this is so staggeringly improbably as to even suggest an answer. Flight must have been a created feature of certain kinds all perfectly coordinated to work together.

Activity GG

Answers should include: glacial cirques, glacial horns, glacial valleys, erratics and striations.

Appendix E – Comprehensive Exam

Comprehensive Exam

Matching – circle the best possible answer:

1. The Father of Modern Geology has been credited to
   a. Charles Darwin
   b. Satan
   c. Charles Lyell
   d. James Hutton

2. The belief that God created but is not now involved:
   a. Theistic evolution
   b. Uniformitariaism
   c. Deism
   d. Naturalism

3. An imaginary construct that geologists use to interpret the rocks:
   a. Geiger counters
   b. The Geologic Column
   c. Radiometric Dating
   d. The Bible

4. Another name for Earth movement:
   a. Tectonics
   b. Earthquakes
   c. Tsunamis
   d. Eruptions

5. "The present is the key to the past":
   a. Geologic reasoning
   b. Chemical analysis
   c. Deism
   d. Uniformitarianism

6. Quartz is high in oxygen and silicon which means it belongs to this family of minerals:
   a. The oxides
   b. The silicates
   c. The borates
   d. The carbonates

7. An example of a light colored volcanic rock would be:
   a. Basalt
   b. Granodiorite
   c. Granite
   d. Rhyolite
8. The Greek word for the "well ordered" universe is:
   a. Geology
   b. Cosmos
   c. Catastrophe
   d. Paleontology
9. An example of a dark colored mineral would be:
   a. Calcite
   b. Potassium feldspar
   c. Sodium feldspar
   d. Olivine
10. The chemical construct used to measure the acidity or base of a substance is called:
    a. A thermometer
    b. A gas meter
    c. The pH scale
    d. The calorimeter
11. The stage of the Flood which buried all land dwelling animals and many sea creatures:
    a. The casual stage
    b. The judgment phase
    c. The burying stage
    d. The Inundatory stage
12. A landform which was cut by water into deep canyons or channels
    a. A water gap
    b. A volcanic gap
    c. A wind gap
    d. A glacial gap
13. An underwater volcano which has had its topped sheared off
    a. A plateau
    b. A gap
    c. A canyon
    d. A seamount

14. An example of a metamorphic rock
    a. Gneiss
    b. Granite
    c. Sandstone
    d. Basalt
15. An example of a plutonic rock
    a. Rhyolite
    b. Granite
    c. Basalt
    d. Limestone
16. An example of volcanic rock
    a. Basalt
    b. Schist
    c. Bytownite
    d. Granodiorite
17. An example of a sedimentary rock
    a. Gneiss
    b. Granite
    c. Shale
    d. Slate
18. One condition necessary for fossilization would be:
    a. Heat
    b. Mutations
    c. Flotation
    d. Rapid burial
19. An example of an Ice Age feature:
    a. Sedimentary rocks
    b. Volcanic eruptions
    c. Striations on rock surfaces
    d. Melting of rocks
20. An example of a volcanic aerosol would be:
    a. Lava
    b. Thunder egg
    c. Hot water
    d. Ash

Appendix F
Suggested Reading and Supplemental Kits
All resources are available from Northwest Treasures.
Kits are made exclusively by Northwest Treasures.
NorthwestRockAndFossil.com

Part I – The History of Modern Geology
- Charles Darwin – The Man and the Myth (Northwest Treasures Kit)
- Genesis and the Age of the Earth (Northwest Treasures Kit)
- The Great Turning Point, Terry Mortenson

Part II – The Origin and Nature of the Earth
- God's Design for Heaven and Earth (Northwest Treasures Kit)
- Creation for Kids (Northwest Treasures Kit)
- Creation Science (Northwest Treasures Kit)
- Biblical Geology for Grades 5-12 (Northwest Treasures Kit)
- Our Created Moon, Don DeYoung and John Whitcomb
- The Young Earth, John Morris
- Geology for Kids (Northwest Treasures Kit)
- Rock Identification for Grades 3-12 (Northwest Treasures Kit)
- The Oceans Book, Masterbooks
- The Weather Book, Michael Oard

Part III – The Genesis Flood
- Geologic Evidence for the Genesis Flood, The Geology of the Colorado Plateau (Northwest Treasures Kit)
- Noah's Flood for Kids (Northwest Treasures Kit)
- Noah's Flood for Grades 3-6 (Northwest Treasures Kit)
- Noah's Flood for Grades 7-Adult (Northwest Treasures Kit)
- The Geology of the Genesis Flood (Northwest Treasures Kit)
- The Complete Earthquakes (Northwest Treasures Kit)
- The Evidence for the Genesis Flood (Northwest Treasures Kit)
- Rock Solid Answers, Edited by Michael Oard
- Grand Canyon National Park Geology (Northwest Treasures Kit)
- Rock Identification for Grades 3-13 (Northwest Treasures Kit)
- Volcanoes for Grades Prek-2nd (Northwest Treasures Kit)
- Yellowstone National Park Geology (Northwest Treasures Kit)

- The Complete Volcano Unit Study (Northwest Treasures Kit)
- Volcanoes and Volcanic Rocks of the Cascades (Northwest Treasures Kit)
- Mt. Rainier and Mt. St. Helens Volcanic Geology (Northwest Treasures Kit)
- Volcanoes, Calderas and Volcanic Rocks of the Western United States (Northwest Treasures Kit)
- Sedimentary, Igneous and Metamorphic Rocks unit studies
- Fossils and the Flood for Grades 5-12 (Northwest Treasures Kit)
- Living Fossils, Dr. Carl Werner
- Evolution: The Grand Experiment, Dr. Carl Werner
- Dinosaur Challenges and Mysteries, Michael Oard
- Dry Bones, Gary Parker
- The Fossil Record, John Morris
- The Fossil Book, Gary Parker
- Dinosaurs for Kids (Northwest Treasures Kit)
- Dragons of the Deep (Northwest Treasures Kit)

Part IV – The Ice Age
- Yosemite National Park Geology (Northwest Treasures Kit)
- Rockhounding Washington (Northwest Treasures Kit)
- Yellowstone National Park Geology (Northwest Treasures Kit)
- Genesis and the Ice Age (Northwest Treasures Kit)
- The Ice Age for Kids (Northwest Treasures Kit)
- The Missoula Flood Controversy, Michael Oard
- The Weather Book, Michael Oard
- Life in the Great Ice Age, Michael Oard
- Uncovering the Mysterious Wholly Mammoth, Michael Oard
- An Ice Age Caused by the Genesis Flood, Michael Oard

All are available from Northwest Treasures, NorthwestRockAndFossil.com, 425-488-6848.

Appendix G
Supplemental Enrichment Activities

Activity 1: How to make a volcano

Time needed: Three separate class periods
Objective

The "how to make a volcano" science project is designed to help young students learn more about earth science by looking specifically at volcanoes. We'll also learn how common household items can be used to build useful models, with an element of creativity required to make the model realistic. Hopefully we'll discover a few new science terms along the way as well. The experiment is done in two steps. First, we figure out how to make a volcano, then we look at fun ways to make it erupt.

Materials needed Make a Volcano with Paper-Mache

-1 newspaper
- 1-2 cups flour, depending on the volcano size desired
- 1-2 cups water
- 1 medium size bowl
- 1 fork or spoon to stir with
- 1 pair of scissors
- 1 roll Scotch or masking tape
- 1 small plastic bag
- 1 pencil or marker

- 1 plastic or glass bottle
- 1 medium size box
- Paint
- 1 medium size paint brush, (a couple more if you have several helpers)
- Rocks, sticks, tips of pine trees or shrubs and anything else you would like to use to decorate the volcano with to make it more realistic.

Some notes on the above materials. First, just about any drink bottle will work, but keep in mind that bottle size will determine volcano height. That's why the amount of flour and water is shown as variable. Second, having sides around the volcano helps keep the "lava" in part 2 of the project contained. However, if having sides is not desired, then substitute a flat piece of cardboard, or even some thin plywood for the box as a stable base for the model volcano. Finally, any paint will do, but a water based acrylic is recommended for easy clean up. They also dry quickly with little need to vent paint fumes. Green, blue, yellow, red, white and black or brown should provide plenty of variety. A suggestion: cut the bottom out of a small Styrofoam glass to use for mixing colors and dispensing the paint.

Preparation

Other than gathering the materials, no advance preparation to make a volcano is needed.

Project Day

It will take at least one class period for them to make a volcano with paper-mache, another to paint and a third to add final decorations and make it erupt. At least one full day will be needed between these steps to allow for drying time.

Project Steps

Get a medium size box and mark where you want to cut the sides.

Cut the box, but do not discard the sides. Place the bottle in the box and draw a circle around its base big enough for the bottle to slip through.

Cut the box sides into about 1 inch strips.

Cut the hole and make sure the box fits over the bottle.

Cover the bottle with a small plastic bag to keep building materials from sticking to the side of the bottle. Make a volcano structure around the bottle with 1 inch cardboard strips that were left over from the cutoff sides of the box. Staples can be used to hold the strips together if desired, but be sure to put plenty of tape around the crater of the volcano, and make sure not to cover the top of the bottle up. If you prefer a stronger structure, chicken wire does great ... but you'll need to supervise that, (and it really isn't needed).

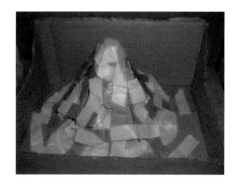

Mix about a cup of flour with enough water to make homemade paste. It should be about the consistency of Elmer's glue. Cut or tear several dozen 1 inch strips of newspaper, but leave at least a sheet or two to put under the box to make the cleanup part easy. Holding one end of a newspaper strip, drag it through the paste and gently squeegee off any excess glue with fingers on your other hand. The goal is for the paper to be wet, but not dripping with glue. Add each glue-soaked strip of newspaper to the volcano support structure, gently smoothing each down as you go. If the forming mountainsides get too much glue on them (you'll know), just add some dry strips to soak it up.

Continue until there are several layers of newspaper strips over the entire mountain, and on the bottom of the box. If you picked a larger bottle, you may need to mix more paste and cut more paper strips to get to this point, but when done, it is time to clean up for the day and let the model volcano dry.

Green makes a great start for grass, trees, etc - and if the volcano is tall, only rocks can be seen near the top. We used brown for that. Sky is blue ... etc. Paint the volcano to make it as realistic as you can. You can go to the next step if desired, but it would be best to let the paint dry first if you can afford the time.

Now decorate the volcano to make it look even more realistic. Add rocks, sticks for fallen tree-trunks, bushes, maybe even houses from a monopoly set, etc. (Then discuss why it might not be a good idea to live near a volcano).

Activity 2: Erupting Volcano

Time needed: varies with method (You will be using the volcano made in Activity 1.)

Baking Soda Methods
There are two general types in this category. Adding soda to liquid or adding liquid to soda.
Vinegar in the bottle first
For erupting volcano projects using this technique, a liquid mixture is put in the bottle first. Baking soda is added through a funnel, or wrapped in tissue paper and forced in the bottle opening.
Trial 1:
1/4 cup vinegar (up to a cup if you have a large bottle)

2 tablespoons baking soda

cherry Jello granules

Place the vinegar in the bottle. This can be done before the bottle is put inside the model volcano in case you want to prepare ahead of time, otherwise, a small funnel works just fine. Stir the baking soda and enough cherry Jello mix to make a pinkish powder. Either wrap the soda mixture in tissue paper or use a funnel to add it directly into the bottle. Tissue helps get all the soda in the vinegar at once, but if the funnel hole is large enough, that method works just fine. Either way, the goal is to get the baking soda into the vinegar as fast as you can. Stand back and watch what happens - Erupting Volcano!
Trial 2:
warm water

1/4 cup vinegar (up to a cup for large bottles)

2 tablespoons baking soda

cherry Jello granules

Fill the bottle about 2/3 full with very warm water. Add vinegar to the bottle. Mix the baking soda and enough Jello together to make the mixture a light pink. When ready for the erupting volcano and add the powder to the vinegar all at once using either the funnel or tissue paper method noted above. The reaction can be quite fast, so add the soda mix quickly and stand back so everyone can see.
Trial 3:
warm water

1/4 cup vinegar (up to a cup for larger bottles)

6 - 8 drops liquid dish soap

2 tablespoons baking soda

cherry Jello granules

Fill the bottle about 2/3 full with very warm water. Add the vinegar and dish soap to the bottle. Mix the baking soda and Jello together until the mixture is pinkish in color. To initiate the erupting volcano project, add the soda mixture to the vinegar all at once using the funnel or tissue paper method. Stand back so everyone can see the erupting volcano.

What's the difference?

Notice that only one item was changed in each of the above trials. Students should understand that controlling the variables in a lab project or science experiment is important to determine what caused any observed changes in the results. The straight vinegar method in **Trial 1** worked just fine for our erupting volcano project. You can use red or orange jello, food coloring, or even kool-aid crystals if desired. We stuck with gelatin crystals because it was easy to use and gave the simulated lava a chunkier texture than food coloring or kool-aid.

Trial 2 had a more violent reaction, meaning it was a bit more spectacular. It was harder to get all the baking soda in the bottle with a funnel before the reaction started producing our lava, but if you're looking for fast reactions, this would be one to consider. In this case the warm water acted as a catalyst, where the temperature of the water helped speed up the reaction. The cherry jello has no noticeable effect on the eruption other than color, and a slight change in the texture of the lava.

For Trial 3, websites differ on what the soap does for the eruption. Yes it changes surface tension, and yes it can make bubbles of its own, but we found no significant change in the frothiness of the lava bubbles or the amount that was produced. It did slow the reaction down a bit and there may have been a bit more lava than with **Trial 1**, but we also didn't have to wait for the bottle to fill with 'lava' before it started spilling out onto our model volcano as in **Trial 1**.

Baking Soda in the bottle first

For ease in doing the experiment, adding the vinegar last wins hands down.

Trial 4:

1/4 cup vinegar (up to a cup if you have a large bottle)

2 tablespoons baking soda

cherry Jello granules

Mix the baking soda and Jello crystals until pinkish in color and use a funnel to get the mixture in the bottle. That's it. When you're ready for the erupting volcano, use a funnel to pour all the vinegar into the bottle at once, then take the funnel out quickly. The key is to get all the vinegar into to bottle as fast as possible and move out of the way. It won't explode, but lava bubbles do ooze out pretty fast once the reaction starts.

Trial 5:

warm water

1/4 cup vinegar (up to a cup for large bottles)

2 tablespoons baking soda

cherry Jello granules

Fill the bottle about 2/3 full with very warm water. Pour in the baking soda and Jello for desired color and mix. It is easier to cap the bottle and shake it outside the volcano, but using a funnel and a straw to mix everything together with the bottle inside the volcano works ok as well. When ready for the erupting volcano, add the vinegar to the bottle with the funnel. Stand back *quickly* so all can see.

Trial 6:

warm water
1/4 cup vinegar (up to a cup for larger bottles)
6 - 8 drops liquid dish soap
2 tablespoons baking soda
cherry Jello granules

Mix enough Jello with the baking soda until you have a slight pink color and add it to the bottle already filled about 2/3rd's full with very warm water. Mix well by shaking the bottle or stirring with a straw. Add the soap and stir gently with a straw so as not to make too many bubbles. When ready for the erupting volcano, add the vinegar quickly through a funnel and move back for others to see.

It is also important to note, the more baking soda and vinegar you use, the more lava will be created.

Diet Coke and Mentos

Trial 7:

12 oz. bottle of regular diet coke
3 mint Mentos

Remove the original bottle and replace it with a freshly opened 12 oz. bottle of regular diet Coke. Flavored diet cokes don't react as well, and neither do flavored Mentos. Original for both is best. With the open bottle in the model, you are ready to go. Now the goal is to get all the Mentos in the bottle as fast as you can. One way that works quite well is to drill a small hole in the center of each mint and hang all 3 by a string. When ready for the erupting volcano, hold the string with all 3 Mentos centered over the bottle hole and let go. Swoosh. The resulting volcanic eruption occurs fast. Drilling holes in candy isn't as easy as it looks, so another method is to use a plastic tube, or roll a small square piece of paper around the Mentos so that you have a small paper cylinder filled with three of the mints. Place another piece of paper under one end of the cylinder so the candy doesn't fall out until you are ready. Rest the bottom paper on the open bottle with the Mentos centered over the hole. When you are ready for the erupting volcano, pull the flat piece of paper out from under the cylinder and let all three mentos fall into the bottle. It may take a little practice getting all the Mentos in the bottle at the same time, but it is a fun way to do the project.

Salt!
Trial 8:
12 oz. bottle of any diet soda

2 tablespoons salt

Any carbonated drink will work, but those with sugar in them slow the reaction down, not to mention that sugar is sticky and makes the cleanup harder. Open a new 12 oz. bottle of diet soda. Flavors may slow the reaction a bit as well, Place the open bottle under the model volcano and you're all set. Put your finger over the end of the funnel and pour about 2 tablespoons of regular table salt into the funnel. When ready for brown lava, hold the funnel over the soda, move your finger off the end and get all the salt in as fast as you can. Then move out of the way. This gets messy as at least 10 of the 12 ounces we tried ended up in the bottom of the model volcano.

What's happening?
With the baking soda and vinegar, a chemical reaction occurs that quickly releases a gas called carbon dioxide, or CO_2. It is the same gas that makes the fizz in a carbonated soda like diet coke, but the way it gets released into a gas is different. To see how it works, we need to look at the components. Baking soda is also known as sodium bicarbonate, (grandma might have called it bicarbonate of soda) and it's a powder that reacts as a slight base. It has a chemical formula $NaHCO_3$. Vinegar is an acid, actually acetic acid in water. It has a chemical formula of $C_2H_4O_2$, or is sometimes written as CH_3COOH to closer represent how it bonds as a molecule. Without getting to deep in the chemistry, when the two are mixed, a hydrogen atom effectively changes places with a sodium atom and we end up with two new substances - carbonic acid and sodium acetate. The carbonic acid then quickly breaks down further into water and carbon dioxide. It is the carbon dioxide escaping from solution that causes the frothy foam we are using as the lava for our erupting volcano. In case you are interested in chemical formulas, ignoring some of the ions, the primary process looks something like this $NaHCO_3 + C_2H_4O_2 \rightarrow NaC_2H_3O_2 + H_2CO_3 \rightarrow NaC_2H_3O_2 + H_2O + CO_2$ soda + vinegar -> sodium acetate + carbonic acid -> carbon dioxide (and some other stuff)

For diet Coke and Mentos, the jury is still out on exactly why the violent reaction occurs. Some say it is strictly mechanical, where the surface of the Mentos provides many cracks and jagged edges as nucleation sites for CO_2 to form larger bubbles and escape. Others claim that the dissolving mint candy interferes with the normal surface tension in the water, which is largely responsible for holding the CO_2 in suspension. Some claim both happen. A quick search on the internet with the seed question "why does diet coke and Mentos explode" will provide all the research needed for the various theories. For our purposes, it is sufficient to say that it makes sense for both the above to occur. The suspended CO_2 in the soda wants to come out of solution all by itself already, and is why if you leave a carbonated soft drink open, it will be flat in just a few hours. The CO_2 stays in solution partly because of the pressure in the bottle, and partly because water molecules tend to stick together. Because of that surface tension, they tend to surround the CO_2 and hold it in suspension. When the Mentos are dropped into the diet soda, just like salt in a carbonated drink, or ice cream in a glass of cola, the CO_2 immediately starts forming bubbles and rises to the surface. It makes sense that the dissolving candy also affects surface tension such that it is easier for the CO_2 to escape. Add lots of spots on the surface of the Mentos where bubbles can start forming and things happen quickly from there. Since the Mentos also sink, and the reaction is quite fast, the escaping gas rapidly moves to the surface, passes it, and takes a bunch (underscore bunch) of the soda with it right out of the bottle. To show that the mechanical effect works all by itself, Trial 8 uses only soda and table salt. The salt dissolves in the water, but there is no chemical reaction to speak of going on. It is an example of a mechanical release of the gas through nucleation, or bubble formation on the edges of the salt crystal themselves. Since the reaction is not as violent, we're thinking the surface tension part of the explanation has merit.

Activity 3: Building a Rock Collection

Time needed: Ongoing project

Introduction: Acquiring rocks, fossils and minerals has been a fun hobby I have enjoyed all my life, from the very first fossil I found to the many fine minerals I have purchased through auctions. Collecting and organizing my precious finds has truly been an enjoyable and learning experience.   Building a nice collection involves time and an increasing desire to learn and organize.   For a more complete study on rock identification, you might be interested in Northwest Treasures' *Rock Identification for grades 3-12*.

Here are some things you can do:
1. I would start at an affordable level. So, start acquiring egg cartons, lots of them. You can paint them or decorate them as you see fit. But one way is to set up a color code system to help you organize and catalogue your various finds. I like egg cartons because the spaces are just the right size for nice specimens.
2. With the rocks you have already collected, go to work sorting them into the various rock types – at least the ones you are able to readily identify. So, for instance, if you have primarily found volcanic rocks, arrange them in ways that make sense to you. Refer to the text to remind yourself of the characteristics of the various rock types.
3. If you are not already familiar with the use of Power Point, this is a great tool for coming up with display grids for your egg cartons. Each of your samples can be recorded in this grid.
4. One of the most overlooked activities in rock collecting is recording the locality where you found your specimen. On the road I use slips of paper to record a general description of the locality and then place it into my bucket of samples. I can always identify the samples later, but I can't always remember the locality.   The locality can be a valuable aid in identifying the specimen at some point in the future.
5. You can continue this procedure with fossils and minerals too. The sky is the limit as to your creativity. Egg cartons will be an easy and affordable way to collect, categorize and store your finds.

You can have some fun learning about soils, minerals and rocks at this web page:
http://www.sciencekids.co.nz/gamesactivities/rockssoils.html

Activity 4: Shake, Rattle and Liquefy

Time needed: less than 60 minutes

Background:

When sediments liquefy, they lose their structure and strength. During earthquake shaking, the individual grains of sand within a deposit collapse on each other. Anything built on them can sink or collapse. Picture a container of balls of slightly different sizes–baseballs, golfballs, marbles. If they were transported by water into the container and then deposited, they would settle with spaces between them. Some of the spaces would be filled with water, some with air. When you shake the container, the balls settle against each other, and the water and air are forced to the surface. That is exactly what happens in a sediment-filled valley. The valley is a large 'container' holding gazillions of 'balls' or grains of sand. Shaking the container simulates an earthquake.

Equipment needed:

Transparent (glass) baking pan
Enough dry sand to fill your pan 1 to 2 inches
A few toy houses or wooden blocks
Water

Purpose:

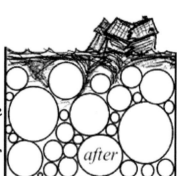

We know that flat river valley bottoms are prone to flooding, but we often think of them as being geologically stable. This experiment will teach you what happens to sandy soils when they liquefy. It will show you how to create a 'model' river valley, then watch how and why houses get damaged or collapse during an earthquake in a seemingly stable geologic environment.

Procedure:

Evenly pour the dry sand into the baking pan.
Mark the level of the sand on the side of the pan.
Place the houses or blocks gently on the surface.
Slowly add water until about two-thirds of the thickness of the sand is saturated.
Gently start shaking the table on which you have placed your baking pan (or the pan itself).

Observations:

You should see the following:
The water will work its way to the surface, flooding the area around the houses,
The houses will start leaning over and sinking into the sand, and
The volume of the sand should decrease by a small amount.

Expanded Activity:

Now be creative. Try the experiment using clay or gravel to separate sand layers and represent different types of sedimentary layers. Watch what happens to the water and the surface of your model of a river valley.

Credits:
Wendy Gerstel, Geologist
Washington Division of Geology and Earth Resources

Activity 5: The Art and Science of Making Fossils

Time needed: about 90 minutes

Background:

Finding plants, animals, and even early humans buried in the geologic strata gives us clues to what our planet Earth was like in the past.

Equipment needed:

Small oven-proof dish or pan

Clay, local, natural source if you're lucky, otherwise play-dough or modeling clay will work (No oil-base clays! They will burn in the oven.)

Leaves, empty shells, dead bugs, etc.

Sand

Purpose:

This experiment will teach you about the process of fossil burial, preservation, and discovery. It will give you the opportunity to think about the types of things (or specimens) one finds buried in sediment, about the sediments and processes that preserve these specimens as fossils.

Procedure:

Layer the bottom of your dish with about one-half inch of the clay.

Explore your backyard or a nearby beach and find things that might become fossilized if they were to be buried, making sure that whatever you pick up is no longer alive!

Next, press your finds gently into the clay. Then, cover this layer of fossils-to-be with a thin layer of sand. This is so your clay layers will part easily after you 'bake' your fossils. Carefully add another layer (or geologic stratum) of clay to your sample. You are now ready to dry your sediments with the buried 'fossils'.

MAKE SURE YOU WORK WITH AN ADULT FOR THIS NEXT STEP. Put the dish in an oven on very low heat. You want to dry your sample slowly so it doesn't crack. This might take an hour or more depending on how wet the clay was. When the sample looks dry, VERY GENTLY remove it from the dish and pry it apart at the sand layer.`

Observations:

You should be able to see:

Your 'fossil' specimens,

The impressions made in the upper and lower clay surfaces, and

How the sample broke along the sand layer.

Expanded Activity:

Look up the difference between 'casts' and 'molds' and see if you can identify each in your sample.

Credits:
Wendy Gerstel and Kitty Reed, Geologists
Washington Division of Geology and Earth Resources

Activity 6: Mudpile Mountain

Time needed: 1-3 hours

The erosion of a mountain by water can be demonstrated by pouring water on a mudpile. This makes it easy for students, even in flat country, to get a better understanding of mountains, their erosional and depositional landforms, and the processes that form them. This outdoor lesson plan is good for groups of any size. Students will construct a 'mountain' of dirt, pour water over it, and identify landforms caused by erosion and deposition.
Caution students not to carry the comparison of mountain to mudpile too far. Mountains are not just piles of dirt but consist chiefly of bedrock. The bedrock is weathered and transported in several different ways and not just simply eroded away by water.

Erosion and Deposition
Weathering breaks up rocks, but what process carries the pieces away? Erosion. It breaks down the Earth's crust and carries away the pieces to deposit them elsewhere. The three main agents of erosion are water, ice, and wind. Running water (rivers and streams) is the most common agent of erosion. Glaciers, immense sheets of moving ice, slowly grind away and carry off large chunks of the Earth's crust. The wind carries sand and dust, which act like sandpaper to wear down the rocks it contacts. **Erosional landforms** illustrated by this activity include streams, canyons, and waterfalls. **Depositional landforms** illustrated include alluvial fans and deltas. This exercise also demonstrates the selective transport of materials by water. Fine particles are carried farther than coarse particles. Fast-moving water carries heavier particles than slow-moving water.

Glossary
Discuss the new vocabulary before starting the activity:

alluvium – silt, sand, clay, gravel, and other loose rock material deposited by flowing water, as in a riverbed or delta.

alluvial fan – the fan-shaped deposit of alluvium left by a stream where it issues from a canyon onto a plain.

bedrock – the solid rock underlying the soil and other loose rock material on the Earth's surface.

canyon – a narrow valley with steep walls, formed by running water.

delta – a triangular alluvial deposit formed where a river enters a large body of water.

deposition – the laying down of material carried by water

erosion – the wearing away of the soil and rock of the Earth's crust

landforms – any physical, recognizable form or feature of the Earth's surface having a characteristic shape and produced by natural causes. Examples include mountains, valleys, deltas, and canyons.

126

sediment – fragmented rock material, such as silt, sand, clay, gravel, carried and deposited by water, wind, or ice.

silt – sediment made up of fine mineral particles smaller than sand and larger than clay.

transport – to carry from one place to another.

weathering – the mechanical and chemical processes by which rock exposed to the weather turns into soil.

Procedure

1. Have each student rule lines 1 centimeter apart across both sides of a popsicle stick, starting from one end. Then have them crayon centimeter-wide bands, neatly, in this order: red, green, orange, blue, yellow, and purple.
2. Take the class outside to a patch of bare ground and dig up dirt, removing larger pebbles and stones. Dump the dirt in a pile and tamp it down to make a 'mountain' about half a meter high.
3. Have the students push their sticks into the 'mountain' and the surrounding 'land', red ends out, so that the sticks are vertical and evenly distributed. (See drawing below) The boundary between the orange and blue bands should be even with the mudpile surface.

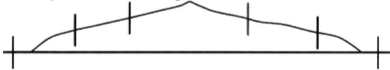

4. Have one student sprinkle the 'mountain' with a watering can so that the 'rain' falls straight down. Let everyone in the class have a chance to be the rainmaker, while the others observe and chart the results.
5. After the 'mountain' is well eroded, ask the following questions:
- Do some markers indicate where erosion is taking place? Where is the material being removed?
- Do some markers indicate where deposition is taking place? Where is dirt being deposited?
- Which of these landforms can you identify: streams, canyons, waterfalls, lakes, deltas, alluvial fans?
- Which is carried farther by the water: sand or silt or rocks? What, if anything, seems to slow erosion on the 'mountain'?

Have the students identify as many of the erosional and depositional landforms on the 'mountain' as they can. Have the students hypothesize why some areas moved at different rates and where the best places to build a house or a waterslide park would be. Continue to erode the 'mountain' and observe it for several days. Measure and record changes in erosion. Draw pictures each day and record measurements in a table. Look for pictures of erosional or depositional landforms from magazines or books.

Activity 7: Do Rocks Last Forever?

Time needed: about 45 minutes, not including freezing time

We think rocks last forever. The boulder we played on in our parents' front yard when we were children is still there for our grandchildren to enjoy. The rock steps to the church are still in use a hundred years later, and the gravestones in the cemetery still mark where our ancestors were laid to rest. These rocks, to us, have lasted forever. But, if you look closely, change is taking place.

This change is called weathering. The term weathering refers to the destructive processes that change the character of rock at or near the Earth's surface. There are two main types of weathering: mechanical and chemical. Processes of mechanical weathering (or physical disintegration) break up rock into smaller pieces but do not change the chemical composition. The most common mechanical weathering processes are frost action and abrasion. The processes of chemical weathering (or rock decomposition) transform rocks and minerals exposed to water and atmospheric gases into new chemical compounds (different rocks and minerals), some of which can be dissolved away. The physical removal of weathered rock by water, ice, or wind is called erosion.

Weathering is not necessarily a long, slow process. In nature, mechanical and chemical weathering typically occur together. Commonly, fractures in rocks are enlarged slowly by frost action or plant growth (as roots pry into the fractures). This action causes more surface area to be exposed to chemical agents. Chemical weathering works along contacts between mineral grains. Crystals that are tightly bound together become looser as weathering products form at their contacts. Mechanical and chemical weathering continue until the rock slowly falls apart into individual grains. *(Patrick's note: Past erosion and abrasion is a matter of interpretation. Present erosion and abrasion is a matter of observation. There is a huge difference. Once we introduce the idea of a global destructive flood into the history of the Earth, then we speed up the erosion and abrasion time.)*

We often think of weathering as destructive and a bad thing because it ruins buildings and statues. However, as rock is destroyed, valuable products are created. The major component of soil is weathered rock. The growth of plants and the production of food is dependent on weathering. Some metallic ores, such as copper and aluminum, are concentrated into economic deposits by weathering. Dissolved products of weathering are carried in solution to the sea, where they nourish marine organisms. And finally, as rocks weather and erode, the sediment eventually becomes rock again – a sedimentary rock.

Two experiments to illustrate the effects of mechanical and chemical weathering are presented below.

Plaster and Ice (Mechanical Weathering)

What you need:
Plaster of Paris, water, a small balloon, two empty pint milk cartons (bottom halves only), a freezer.

What to do:
- Fill the balloon with water until it is the size of a ping-pong ball. Tie a knot at the end.
- Mix water with plaster of Paris until the mixture is as thick as yogurt. Pour half of the plaster in one milk carton and the other half in the other.
- Push the balloon down into the plaster in one carton until it is about $\frac{1}{4}$ inch under the surface. Hold the balloon there until the plaster sets enough so that the balloon doesn't rise to the surface.
- Let the plaster harden for about 1 hour.
- Put both milk cartons in the freezer overnight.
- Remove the containers the next day to see what happened.

What to think about:
What happened to the plaster that contained the balloon? What happened to the plaster that had no balloon? Why is there a difference? Which carton acted as a control? Why? How does this experiment show what happens when water seeps into a crack in a rock and freezes?

What should have happened:
The plaster containing the balloon should have cracked as the water in the balloon froze and expanded. Explain that when the water seeps into cracks in rocks and freezes, it can eventually break rocks apart.

A Sour Trick (Chemical Weathering)

What you need:
- Lemon juice, vinegar, medicine droppers, two pieces each of limestone, calcite, chalk, and quartz. Extra samples may be purchased from *Northwest Treasures*.

What to do:
- Put a few drops of lemon juice on one piece of each of the four rock types.
- Put a few drops of vinegar on the other piece of each of the four types.
- Look and listen carefully each time you add the lemon juice or the vinegar.

What to think about:

What happens when you put lemon juice on each rock? What happens when you put vinegar on each rock? Did the lemon juice and vinegar act the same way on each rock? Why did some of the rocks react differently? What does this experiment have to do with weathering?

What should have happened:

Lemon juice and vinegar are both weak acids. The lemon juice contains citric acid and the vinegar contains acetic acid. These mild acids can dissolve rocks that contain calcium carbonate. The lemon juice and vinegar should have bubbled or fizzed on the limestone, calcite, and chalk, which all contain calcium carbonate. There should not have been a reaction on the quartz, which does not contain calcium carbonate. Explain that water commonly contains weak acids that dissolve rocks containing calcium carbonate and other minerals.

Activity 8: Everyone Needs a Rock

Time Needed: about 1 hour depending on where you go to look for rocks

Background:

Throughout time, societies have valued rocks and mineral resources in their natural state. The great stone circle at Stonehenge, England, is a famous example that is still controversial. Historians are still not sure what peoples placed them there and for what reasons. The monolithic stone figures on Easter Island were carved by unknown ancients. Each year thousands of people travel to see the naturally occurring stone features in the western United States at Devil's Tower, Yellowstone National Park, Bryce Canyon, Zion National Park, the Grand Canyon, and Yosemite National Park. Today as always, people have strong feelings about rocks and minerals. There are museums with remarkable rock collections and other geologic specimens in many cities and at most major universities. There are also rocks and mineral resources to be appreciated in almost every neighborhood. The first step in discovering some of these natural wonders is to look and observe.

Activity:

Ask the students the following questions: "What are five ways we use rocks?" "Have you made a rock creation: if so, describe it."

Discuss with the students where rocks come from and why they are important. (Include things like rock for roads, foundations of houses, skateboard parks, bauxite for aluminum cans, gold and silver for fillings...)

Read: *Everybody Needs a Rock*. *Patrick's note: This book and the list of books at the end of the activity will be evolutionary based. Try your hand at attempting to identify the underlying evolutionary assumptions using the knowledge you have gained in studying through Bedrock Geology.*

Materials:

1. The book, *Everybody Needs a Rock* by Byrd Baylor (available at the library, or for sale through many bookstores).
2. Paper and pencil for each student.

Have each student write or tell his or her own personal rule #11, based on the story.

Take the class rock hunting. Rules 1 through 10 must be used. The students will learn these as you read the story. A review may be needed. Be sure to remind them to use their own rule #11. This could be done as a field trip to a canyon or a park or a walk on the school grounds. Perform the "smelling test" as outlined in the story. Have the students share with the class where their rock came from.

Check for understanding:
- Have the students make up their own rock game, "I happen to have a rock here in my hand..."
- Have the students write a story about where their rock came from. Have them illustrate their stories.

Vocabulary:

Rockhound – one who hunts and collects gemstones or minerals (such as quartz, agate or petrified wood) as a hobby; an amateur mineralogist.

Lesson Extension:

By examining the colors in the rock, students may also want to investigate the mineral content in their rock. (Provide copies of Field Guides to Rocks and Minerals.) Over the next month, keep a list of any rocks you find in your neighborhood. (Examples: Pebbles in the driveway cement, rock walls around the neighbor's garden, marble in an important building, a piece of artwork in the park, etc.) List the location the rock was found; a description of the rock (color, size, texture); what the type of rock is if you know or can find out (does a name plaque give the name of the mineral?). You may want to consult a rock identification book for help. (Simon & Schuster's Guide to Rocks and Minerals, Philip's Minerals, Rocks and Fossils, or the Peterson Field Guide: Rocks and Minerals are a few of the good reference books with helpful photos and descriptions. Every library has these or similar books.)

Every student's list should answer the questions:
- K-3 Where did you see the rock? What color was it? Was it smooth or rough? Was it dull or shiny? How big was it?
- 4-12 For each rock type found, list the location and how it was being used (driveway for paving, bank building foundation, city park sculpture, etc.). Give the color, size, texture. If the rock is identified on a sign or plaque, include the information. (Example: There is a large piece of green serpentine in the downtown Olympia Timberland Library in which a seal is sculpted. According to the label, the serpentine is most probably from Jade Creek, near Mount Vernon, Washington).

References and Suggested Readings:
- Baylor, Byrd; Parnall, Peter, illustrator, 1974, Everybody needs a rock: Scribner, 32 p.
- Bishop, A. C.; Woolley, A. R.; Hamilton, W. R., 1999, Philip's minerals, rocks and fossils; 2nd ed.: G. Philip, in association with the Natural History Museum, 336 p.
- Knoblach, D. A., 1994, Guide to geologic, mineral, fossil, and mining history displays in Washington: Washington Geology, v. 22, no. 4, p. 11-17.

- Pough, F. H., 1996, A field guide to rocks and minerals; 5th ed.: Houghton Mifflin Co. (Peterson Field Guide 7), 396 p.
- Prinz, Martin; Harlow, George; Peters, Joseph, editors, 1978, Simon& Schuster's guide to rocks and minerals: Simon and Schuster, 607 p.
- Waugh, Peter; Sunderland, Don, illustrator, 2002, The great Cannon Beach mouse caper: Educare Press, 99 p.

Activity 9: Rock Types Activities

Sedimentary rocks activity – Sedimentary rock can form out of soil, silt, sand, seashells, and bits of pieces of everything. The tiny bits settle and build up in thick layers. Minerals from water seeps into the spaces between the particles and bond them into solid rock. The Grand Canyon is made up of sandstone, limestone, and shale which are all sedimentary rocks. **For this project you will need:**

- Mixing bowl
- Sand
- Epsom salt
- Spoon
- Small paper cup
- Water

Mix 1 cup of water and 1/2 cup of Epsom salt in a mixing bowl. Stir until most of the salt dissolves. Place about 1 inch of sand in a small paper cup. Add enough salt solution to just cover the sand. Mix well. Let the mixture stand until dry (about 2 to 3 days). Cut off the paper. Take a close look at the rock. Where is the salt? How does it hold the sand together?

Metamorphic rocks activity – Metamorphic rock used to be igneous or sedimentary rock that was changed by heat or pressure or both. The Appalachian Mountains including the Blue Ridge are mainly metamorphic rock with a little igneous rock. The same is true of the mountains around Mt. Baker, WA. **For this project you will need:**

- Different colors of Play Dough or Modeling clay
- Waxed paper
- 3 books

Make about 2 dozen pea-sized balls from clay of different colors to represent rock particles. Place the clay balls close together on a piece of waxed paper. Place a second sheet of waxed paper on top of the clay balls. Now, put stack some books on top of the waxed paper. Imagine that the books are layers of rock building up on top of the rock particles. The pressure of the rock particles increases as you add each book. Heat builds as the rock particles are pushed deeper into the Earth's crust. Now remove the books and peel away the waxed paper. Look at the clay. An entirely new kind of rock has been formed. The same thing happens to igneous and sedimentary rocks when they are changed.

Igneous rocks activity – Igneous rock was once hot liquid inside the earth. Rock that erupted from volcanoes and cooled on the surface is called **extrusive** igneous rock. Igneous rock that cooled more slowly inside the Earth is called **intrusive**. Apache Tear and the Hawaiian Islands are mainly igneous rock. Much of the Cascade Range is igneous rock. **For this project you will need:**

- Spray cooking oil
- Waxed paper
- Hot plate or oven range burner
- Small plastic cup
- Plastic spoon
- A cooking pan
- Wooden spoon
- 1 bag of large marshmallows
- 6 tablespoons of butter
- 6 cups of Rice Krispies

These ingredients will form the molten rock. Place the marshmallows and butter in a pan. Stir and melt. Add the Rice Krispies. Stir. Place a cup of the warm, Rice Krispy mixture on the waxed paper square. Make sure your hands are sprayed with cooking oil. Sprinkle and roll into the mixture other ingredients like raisins, chocolate chips, nuts, M&M's with your hands. These ingredients will form the rock surface. After it cools and hardens, you have an edible igneous rock!

Activity 10: Creeping Crystals

When you start work on a creeping crystals science project, be careful. You may feel like a mad scientist when you see the results!

Create some creeping crystals

What You'll Need:
- Saucepan
- Water
- Measuring cup and spoon
- Epsom salts
- Mixing spoon
- Green food coloring
- Metal bottle caps
- Pie pan
- Strings
- Pennies

Step 1: Heat 1/2 cup water in the saucepan over medium heat. Measure 1/2 cup Epsom salts, and mix salts in hot water by spoonfuls until no more dissolves. Turn off heat.

Step 2: Add 2 drops green food coloring, and stir. Let mixture cool for 20 minutes or more.

Step 3: Arrange the bottle caps in the pie pan. Carefully pour the warm solution into the bottle caps until the caps are full.

Step 4: Place strings in the caps. Use pennies to weigh down the strings so they can't float.

Step 5: Place the pan where the water will evaporate quickly. Warm areas with good airflow increase evaporation.

Step 6: Watch the caps for a few weeks for crystal growth. If a crust forms on the water's surface, use a pencil to make a hole so you can watch the crystals form.

What Happened? Epsom salts (magnesium sulfate) dissolve in water. Heating the water allows more salt to dissolve into the water. When you do this, you create what chemists call a supersaturated solution. After the water is poured into the caps, it begins to cool. Now the water can't hold as much salt, and crystals begin to form. As the water evaporates, the crystals continue growing. *Look closely and compare a few grains of table salt and Epsom salt.* You can see the crystals are different shapes. When large crystals grow, they are made out of the same repeated shapes.

Appendix H
Quizzes Answer Key

Part I – The History of Modern Geology
Section 1 and 2
1. Charles Lyell
2. James Hutton
3. Nicolas Steno
4. The Enlightenment
5. Naturalism
6. Uniformitarianism
7. Deism is a belief that there is a God who created all things, but He is not involved in its maintenance. Therefore miracles are not possible.
8. The Geologic Time Table or Column

Part II – The Origin and Nature of the Earth
Section 1
1. That the universe had a beginning and that someone or something greater than the universe brought it into existence.
2. Day 1 – the space and the Earth, light; Day 2 – atmosphere; Day 3 – waters gathered into one place and the appearance of dry land, vegetation on Earth; Day 4 – stars, sun and moon; stars placed into the space; Day 5 – Waters teem with living things, birds; Day 6 – land-dwelling animals and man
3. A framework is the basic outline or construction upon which the rest of the "house" is built; it is keeps the house from collapsing; Genesis is a dependable framework because it is revelation from God and no one was around to record what happened. Therefore we cannot know in and of ourselves what took place in the past without God revealing it to us.
4. The DNA in living things involves very ordered and complicated information that must all work together or it does not work at all. The possibility of all these various functioning parts coming into existence without the aid of a Creator is not possible.

Section 2
1. Cosmology. It means the "well-ordered" universe. The Greeks saw that the universe was a product of an intelligent designer.
2. Silicon and oxygen.
3. The proton, the neutron and the electron. The nucleus holds the protons and neutrons and it is what the electrons orbit.
4. He was created in the image of God.
5. Could include quartz and potassium feldspar. Could include iron and hornblende.
6. Quartz, potassium feldspar, sodium feldspar, calcium feldspar, olivine, calcite, hornblende (amphibole), augite (pyroxene), biotite mica, muscovite mica, jasper (a type of quartz) and magnetite (iron oxide)
7. Silicates, carbonates, sulfides, sulfates, phosphates, halides, oxides, borates, native elements

Section 3
1. Hydrogen and oxygen
2. Water
3. 70%. It purifies and carries nutrients and drinking water.
4. The pH scale measures the acidity or base of Earth's substances on a scale from 0 to 14, 0 being the most acidic and 14 being the most basic. Bases and acids are necessary for life. Water neutralizes these and so balances life's delicate needs. This delicate balance is obviously a product of a caring and loving God who is all wise.

Section 4
1. Oxygen and nitrogen
2. It keeps out the sun's harmful radiation and effects and it protects us from atmospheric radiation.
3. It further protects us from harmful radiation of space.

Part III – The Genesis Flood
Section 1
1. Could include the fountains of the great deep bursting open, the inundation of the Earth by water, the covering of all mountains on Earth, the killing of all land-dwelling life on Earth, the receding flood waters, the reproduction of life from life that was on the Ark.

2. The Bible is a revelation from God that tells us not only what is important, but what took place historically. So, it is a historical record as well.

Section 2
1. The inundatory and the receding stages
2. The inundatory – massive graveyard fossils; receding – Grand Canyon

Section 3
1. Folded sedimentary rocks had to have been laid down in quick succession and while still pliable, bent. This is because all the layers are bent in the same direction; Devil's Tower is a remnant of a volcano having had its top and outside weaker sediments removed during the receding phase of the Flood; Planations surfaces are large flat surfaces evenly planed by some force other than gradual erosion. These would have been produced during the sheet flow or receding stage of the Flood; seamounts would have been like Devil's Tower, only under water.

Section 4
1. Basalt and rhyolite
2. Gneiss and schist
3. Limestone and sandstone
4. Basalt and rhyolite are produced in volcanic eruptions or flow and contain either light or dark colored minerals. Metamorphic rocks have been changed it is thought, by heat and pressure to rearrange the mineral composition. Sedimentary rocks have been laid down by water or mud and then hardened by chemical processes not in operation today.
5. Clast is the Greek word for "broken". Clastic sedimentary rocks are those which contain bits and pieces of other rocks and minerals glued together by minerals such as calcite or quartz.
6. Chemical sedimentary rocks are those which have been produced by chemical processes not now operating on the scale observed in the rock record.
7. The abundance of limestone all over the Earth would speak of a massive amount of water and chemical processes not now operating.

Section 5

1. Rapid burial to preserve from decay and the right chemical process to quickly harden the plant or animal into stone.

2. The fossil record shows that life arose quickly and diversified into varieties within kinds. There are no stages of evolutionary development recorded in the fossil record.

3. A transitional fossil would be a life-form on its way to becoming another kind of creature or plant. There should be perhaps millions of these preserved to some degree in the fossil record if evolution were true. In contrast Genesis teaches that God created distinct kinds of life which had the capability to produce variations but within its own kind.

4. Could include the following: (1) life arose from non-life; (2) the basic cell evolved into invertebrate animals; (3) invertebrate animals developed into vertebrate animals (fish); (4) fish developed into land animals; (5) land animals (amphibians) developed into reptiles; (6) reptiles developed into mammals and (7) mammals developed into man. In the Genesis order, the first life consisted of plants. Then life appeared in the ocean. Then life appeared in the skies as birds. Life then appeared on land and in the same time period as man. Not only is the order different, but life was created according to kinds with no transition to other kinds and with the capacity to reproduce according to their kinds.

Part IV – The Ice Age
Section 1

1. Could include: Glacial cirques; glacial valleys; erets; horns; moraines

2. Loess is very fine wind-blown dust. With cooling temperatures toward the end of the ice age, a great deal of wind would have been present stripping the soil of dust and quickly burying things such as mammoths. This ability was demonstrated during the American Dust Bowl.

Section 2

1. The ice age would have had to fit between the end of the Flood and the advent of Moses. Conditions brought about by the Flood could not have persisted indefinitely as the cause would have ended as Genesis stated within a year.

2. The rapid build up of ice and snow and the rapid catastrophic melting of the ice and snow laid down in the initial stages.

3. The Genesis Flood combines the increase in volcanic aerosols brought about by the tectonic activity of the initial stages of the Flood combines with the warmer oceans produced by the release of magma.
4. The two main ingredients would be a cooler atmosphere and warmer oceans. Both of these were uniquely combined as a result of the Genesis Flood.

Polished ammonite fossil from Morocco

Appendix I – Comprehensive Exam Answer Key

Answers to the Exam

1. d
2. c
3. b
4. a
5. d
6. b
7. d
8. b
9. d
10. c
11. d
12. a
13. d
14. a
15. b
16. a
17. c
18. d
19. c
20. d

Made in the USA
San Bernardino, CA
01 February 2014